西电科技专著系列丛书

U0170237

# 精选科技英语写作典型错误分类解析

## A Classified Analysis of Select Typical Mistakes in Scientific Writing

秦荻辉　编著

西安电子科技大学出版社

# 内 容 简 介

随着全球化的不断深入,我国大学生、研究生、专业教师和科技人员用英语撰写论文的需求越来越迫切。为了满足这种需求,本书作者收集了出现在真实论文中的大量典型错误实例,对其进行了分类和详细的分析,期望能帮助读者在撰写英文论文时少犯(或不犯)类似的错误,从而提高论文的英语表述质量。

本书适合各类大学本科高年级学生、硕士生、博士生以及专业教师和广大科技工作者阅读参考。

**图书在版编目(CIP)数据**

精选科技英语写作典型错误分类解析/秦荻辉编著. —西安:西安电子科技大学出版社,2021.3

ISBN 978 - 7 - 5606 - 5987 - 9

Ⅰ.①精… Ⅱ.①秦… Ⅲ.①科学技术—英语—写作—高等学校—教学参考资料 Ⅳ.①N43

中国版本图书馆 CIP 数据核字(2021)第 019109 号

策划编辑 马乐惠
责任编辑 卢杨 阎彬
出版发行 西安电子科技大学出版社(西安市太白南路 2 号)
电 话 (029)88242885 88201467 邮 编 710071
网 址 www.xduph.com 电子邮箱 xdupfxb001@163.com
经 销 新华书店
印刷单位 咸阳华盛印务有限责任公司
版 次 2021 年 3 月第 1 版 2021 年 3 月第 1 次印刷
开 本 787 毫米×1092 毫米 1/16 印张 13
字 数 232 千字
印 数 1~2000 册
定 价 29.00 元

ISBN 978 - 7 - 5606 - 5987 - 9/N

XDUP 6289001 - 1

* * * 如有印装问题可调换 * * *

# 前　　言

　　现在我国大学生的毕业论文的文摘均要求附有英文译文，向科技杂志提交的论文也需附有英文文摘，研究生（特别是博士生）、专业教师及科技人员在国外杂志上发表论文的要求也越来越迫切。但总体上说，我国不少研究生、专业教师和科技工作者用英语撰写学术论文的能力还比较薄弱，对英美科技人员偏爱的句型、固定表示法、词汇搭配及一些语法条文并不熟悉，因而投到国外科技杂志的论文常常被退回来要求修改或重写。评阅人的典型意见有："You should revise the manuscript with particular emphasis on the use of English." "The use of the English language is not good enough to allow acceptance of the paper." "Your English is bad. Not only the wording，but sentence structures are sometimes hard to follow. IT MUST be reedited seriously."等。

　　本书编者三十多年来一直在校阅修改《西安电子科技大学学报》论文的英文文摘、研究生和教师参加国际会议的论文及向国外科技杂志投稿的文章，在此过程中摘录了大量常见典型错句，本书就是在对这些错误进行分类详解的基础上编写而成的。希望读者通过学习本书能够在今后的英文写作中少犯（或不犯）类似的错误，从而提高学术论文的英语表达质量。

　　科技英语写作的基础是科技英语语法知识及英美科技人员惯用的一些句型，为此读者可以参阅由本书编者所著的以下三本著作：(1)由北京外语教育和研究出版社出版的"普通高等教育'十一五'国家级规划教材"——《科技英语语法》(被教育部评为"精品教材"，向各高校推荐使用)；(2)由西安电子科技大学出版社出版的《科技英语写作高级教程(第三版)》；(3)由西安电子科技大学出版社出版的《学术论文写作英语惯用法》。这些书中的例句均摘自各种英美科技书刊，颇具典型性，其中的大量句型在撰写学术论文时可直接套用。

　　本书适合各类大学本科高年级学生、硕士生、博士生、专业教师和广大科技工作者阅读参考。

编　者

2020 年 11 月

于西安电子科技大学

# 目　　录

# 第一部分　词汇问题

## Ⅰ、冠词

### 1. 冠词的丢失(这一错误是最常见的)

例 1

【汉语原句】　这个方法的误差在信噪比(SNR)低的时候比多级方法的小。

【英语错句】　The error of this method is smaller than the multistage method at low SNR.

【错误分析】　(1)"of"应该改为"(caused) by"。(2)比较对象不一致,在"than"后应该加上"that by"。(3)"at"应该改为"for"为好。(4)"SNR"是个数值,所以它是可数名词单数,因此,在"low"之前应该加有"a"。

【改正后的句子】　**The error by this method is smaller than that by the multistage method for a low SNR.**

例 2

【汉语原句】　映像可以表示为 $g = T(f)$,式中 $f$ 是输入图像的灰色数值。

【英语错句】　The map can be expressed as

$$g = T(f)$$

where $f$ is the gray value of input image.

【错误分析】　"input image"是一个可数名词,这里它是单数,所以其前面应该有冠词"an"或"the"。

【改正后的句子】　**The map can be expressed as**

$$g = T(f)$$

**where $f$ is the gray value of an/the input image.**

例 3

【汉语原句】　目标的运动状态对扩展式卡尔曼滤波器(EKF = extended

Kalman filter)的跟踪精度的影响是非常明显的。

【英语错句】　The influence of moving state of the target is very strong for the tracking accuracy of EKF.

【错误分析】　(1) 在"词汇搭配"方面，由于句中有名词"influence"，所以应该使用"the influence [effect；impact] of A on [upon] B"搭配模式。(2) 这里与该名词"influence"搭配的形容词通常是"great"。(3) 由于可数名词"moving state"有后置定语"of the target"，所以其前面要有定冠词；"EKF"是个特定的具体的东西，因此在其前面也应该加上一个定冠词。

【改正后的句子】　The influence of the moving state of the target on/upon the tracking accuracy of the EKF is great.

### 例 4

【汉语原句】　用户公钥是他的身份信息，例如邮箱地址、IP 地址、电话号码等。

【英语错句】　User's public key is his identity information，such as，e-mail address、IP address、telephone number and et al.

【错误分析】　(1) 由于这里"user"是单数，所以在"user's"前要加定冠词。(2) "his identity information"最好改为"the information on his[her] identity"。(3) 在"such as"后的逗号要去掉。(4) 在英语中是没有顿号的，所以要把它改成逗号。(5) "and et al."应改成"，etc."。("et al."只用于人)，在"etc."前是不能加"and"的。

【改正后的句子】　The user's public key is the information on his/her identity，such as e-mail address，IP address，telephone number，etc.

### 例 5

【汉语原句】　本文对区间双正交小波所具有的多尺度边缘提取能力进行了理论分析。

【英语错句】　This paper theoretically analyzes the multi-scale edge detection ability of interval biorthogonal wavelet.

【错误分析】　(1) 本句的主要错误在于有关"ability"的句型不对，原作者只是根据汉语把词汇堆在了一起。正确的句型是"the ability of A to do B"。(2) 由于"wavelet"是可数名词单数形式，所以在其前面应该有冠词。(3) 英国和美国(后简称英美)科技人员喜欢用名词来表示动作，因此最好采用"make an analysis of …"这一表达方法。

**【改正后的句子】** This paper makes a theoretical analysis of the ability of the interval biorthogonal wavelet to detect the multi-scale edge.

例 6

**【汉语原句】** 由该模块处理的信息，通过数据缓冲器被送到 DAC，在那儿数字图像数据被转回成模拟图像信号。

**【英语错句】** The image data processed by this module is sent to DAC through data buffer. There the digital image data are transformed back to analog image signal.

**【错误分析】** (1) 一般情况下，"data"是用作为复数的，所以"is"应该改写成"are"。(2) 由于"buffer"和"signal"均为可数名词单数，所以在它们的形容词之前应该分别加有冠词"a"和"an"。(3) 在"DAC"之前应该加上定冠词。(4) 为了使句子显得紧凑，应该把两个句子合并成一句，把第二个句子变成一个非限制性定语从句，采用关系副词"where"连接。

**【改正后的句子】** The image data processed by this module are sent to the DAC through a data buffer, where the digital image data are transformed back to an analog image signal.

例 7

**【汉语原句】** 在 11～13 页上，作者描述了自从 20 世纪 80 年代以来电子学方面我们所取得的成就。

**【英语错句】** On p. 11－13, the author describes the achievements we obtained since 1980's.

**【错误分析】** (1) 当页码为复数时，一定要用"pp."来表示"pages"的意思，不少人错误地惯用"p."来表示。(2) 凡句中有"since"引出的短语或时间状语从句时，主句一定要用完成时态。(3) 与"achievement"搭配的动词"取得"应该是"make"。(4) 表示"年代"一定要有两个标志——前面有定冠词"the"和尾部有"s"，而"'"则可有可无。

**【改正后的句子】** On pp. 11－13, the author describes the achievements we have made since the 1980's.

例 8

**【汉语原句】** 其中之一称为比较组，而这里所用的测量是测试 A 和 B。

**【英语错句】** One of them is called as comparison group, and measurements used here are Test A and B.

【错误分析】　(1)表示"把 A 称为 B"的表达法有好几种，其中有的要带有"as"，而有的则是不带"as"的，读者一定要把它们记清楚。常见的有以下一些（在此用被动式给出）：

A is called B.

A is named B.

A is termed B.

A is known **as** B.

A is referred to **as** B.

A is spoken of **as** B；

(2)由于"group"是可数名词单数，所以其前面一定要有冠词，在此应该用定冠词。(3)由于"measurements"后面带有后置定语"used here"，所以其前面一般应该加上定冠词。(4)测试涉及两个，所以一定要用复数"Tests"，这也是某些读者常犯的一个错误。又如：

Figs. 1－1 and 1－2　　　图 1－1 和 1－2

Eqs. (2－5)to(2－8)　　　式(2－5)到(2－8)

Chaps. 3 and 4　　　　　第 3、4 章

【改正后的句子】　**One of them is called the comparison group，and the measurements used here are Tests A and B.**

## 例 9

【汉语原句】　离线第三方不必完全可信。

【英语错句】　Off-third party needn't to be completely trusted.

【错误分析】　(1)"party"是可数名词单数，所以其前面一定要有冠词，根据句意，这里应该使用定冠词。(2)"needn't"在科技文中应该分开来写成"need not"。(3)由于用了"need not"，那"need"就是情态动词了，所以其后不能跟有带"to"的不定式了。（如果把"need"看成是实意动词，则应该用"does not need to be completely trusted"。）

【改正后的句子】　**The off-third party need not be completely trusted.**

## 例 10

【汉语原句】　这种生成器产生的序列的周期几乎都大于任意给定的值。

【英语错句】　Almost all of the periods of sequences generated by such generator are greater than arbitrary given value.

【错误分析】　（1）在"such"后跟有可数名词单数时要用"不定冠词＋单数名词"。（2）由于"value"为可数名词单数，所以其前面一定要有不定冠词"an"（因为后面一个词以元音开头）。（3）"任意"在此修饰"给定的"，所以应该使用副词而不该用形容词，因此要把"arbitrary"改成"arbitrarily"。

【改正后的句子】　**Almost all the periods of sequences generated by such a generator are greater than an/any arbitrarily given value.**

### 例 11

【汉语原句】　Hash 算法是可以把任何长度的数据压缩成具有固定长度数据的一种方法。

【英语错句】　Hash algorithm is a scheme that data of any length are compressed into data of a fixed length.

【错误分析】　（1）在"人名＋名词"前，要加定冠词。（2）这里用"that"引导同位语从句是错误的，只能用定语从句，所以应该把"that"改写成"by which"。（3）由于两个复数形式的"data"后均有定语"of … length"，所以它们之前加上定冠词为好。

【改正后的句子】　**The Hash algorithm is a scheme by which the data of any length can be compressed into the data of a fixed length.**

### 例 12

【汉语原句】　这能够以尽可能低的收缩频率来满足系统的输入和输出速率。

【英语错句】　This can satisfy system's input and output rate with as low as possible systolic frequency.

【错误分析】　（1）在"system's"之前要加一个定冠词"the"。（2）"尽可能低的频率"应该译成"as low **a** systolic frequency as possible"，在 low 之后要加上一个不定冠词"a"，因为"frequency"是可数名词单数，其前面一定要有冠词，这里由于是"as…as"句型的关系，所以要采用不定冠词的特殊位置。

【改正后的句子】　**This can satisfy the system's input and output rates with as low a systolic frequency as possible.**

### 例 13

【汉语原句】　最后证明了这一实验系统可以通过因特网进行遥控。

【英语错句】　In the end，it proves that this experiment system can be remotely controlled through Internet.

**【错误分析】**　（1）表示"最后"之意，绝大多数情况使用"finally"一词，指"最后一点"；也有个别情况用"last"或"lastly"的。但决不能用"at last"或"in the end"，那是表示"最终，终于"之意。（2）动词"证明了"由于没有指出动作的执行者，所以应该使用被动语态，即"it is proved [shown]"。（3）"experiment"应该改为形容词"experimental"。（4）"remotely controlled"一般用"remote-controlled"表示。（5）在"Internet"前一般有定冠词。

**【改正后的句子】**　**Finally，it is shown that this experimental system can be remote-controlled through the Internet.**

## 2. 冠词位置不对

**例 14**

**【汉语原句】**　该系统能获得与非扩频系统一样高的频带效率。

**【英语错句】**　The system can obtain a band efficiency as high as a non-spectrum-spreading system does.

**【错误分析】**　本句的错误在于"as high as a non-spectrum-spreading system does"这一部分，这是中国理工科学生和科技人员经常犯的错误，其主要原因在于认为"as high as …"为形容词短语作后置定语修饰"a band system"。问题是凡是形容词短语作后置定语均可扩展成一个定语从句，这时错误就显现出来了，即"… a band efficiency that is as high as a non-spectrum-spreading system does"，这里关系代词"that"就代替了"a band efficiency"，这样就可以看出在从句中的比较对象不一致了，一个是"频带效率"，而另一个是"非扩频系统"，这两者是不能比较的，若要比较的话，则要把"a non-spectrum-spreading system does"改写成"that of a non-spectrum-spreading system（is）"，这样从语法上讲是正确的，但英美科技人员一般是不用这种表达方式的，他们喜欢把"as … as "句型套在名词"a band efficiency"上，这样句子显得简洁，但这时一定要注意不定冠词的特殊位置，即"as high **a** band efficiency as…"。

**【改正后的句子】**　**The system can obtain as high a band efficiency as a non-spectrum-spreading system does.**

**例 15**

**【汉语原句】**　这个信号太小了以至于不能使晶体管 Q4 导通。

**【英语错句】**　This is a too small signal that it cannot make transistor Q4 conduct.

【错误分析】 （1）主句用了名词短语作表语，这一点是可取的，因为英美科技人员比较喜欢用名词作表语的结构（当然也可以用形容词作表语，这里可写成"This signal is too small …"），问题在于不定冠词"a"的位置放错了，这时一定要放在"small"之后，也就是"This is too small a signal …"。（2）由于本句使用了"too"，所以后面不该使用由"that"引导的结果状语从句，而应该使用"too…to（do）"句型，而且应该使用"不定式复合结构"的形式。

【改正后的句子】 **This is too small a signal for transistor Q4 to conduct（或：for transistor Q4 to be on）.**

例 16

【汉语原句】 必须了解这种材料是否能经受住这么大的力。

【英语错句】 It is necessary to know if this material can stand a so large force or not.

【错误分析】 （1）根据传统语法规则"if"一般不能与"or not"连用，所以应把"or not"去掉。若想保留它的话，则应把"if"换成"whether"。（2）不定冠词"a"的位置错了，由于有"so"修饰了"large"，所以必须把"a"放在"force"之前。

【改正后的句子】 **It is necessary to know whether this material can stand so large a force or not.**

例 17

【汉语原句】 若压力太大就会使该器皿破裂。

【英语错句】 A too large pressure will break the vessel.

【错误分析】 本句用了一个名词短语来表示条件是对的，这种方式是英美人偏爱的写法，但可惜的是冠词的位置错了，在这里应该使用特殊位置，即"Too large a pressure"。

【改正后的句子】 **Too large a pressure will break the vessel.**

## 3. 误用冠词或用错冠词

例 18

【汉语原句】 我们能够充分利用地球的信息。

【英语错句】 We can make a full use of the informations of earth.

【错误分析】 （1）词组"make use of"加上"full"后其前面是不加冠词"a"

的。而词组"参加(take part in ⋯)"若加上"积极的(active)"后则要加上冠词"an",成为"take an active part in ⋯(积极参加⋯⋯)"。(2)"information"是一个不可数名词,要表示"许多信息"的话,可以使用"much[a lot of;a great deal of] information"来表示,也可以用"many pieces of information"来表示。具有类似用法的名词有"literature(文献)""intelligence(情报)""knowledge(知识)"等。(3)"information"后面不能跟"of"而应该跟"on""about",这属于词汇的搭配问题。(4)在"earth"前一定要有定冠词"the"。

**【改正后的句子】**　**We can make full use of the information on/about the earth.**

**例 19**

**【汉语原句】**　在第二部分提供了"Yes"和"No"两种选择。

**【英语错句】**　In the Part II are provided with two choices "Yes" and "No".

**【错误分析】**　(1)在表示"第几"的"Part II"这类表达形式之前是不能有冠词的。又如:"Chapter 1(第一章)""Sec. 4(第四节)""Eq. 1 - 5(式 1 - 5)""Fig. 5 - 1(图 5 - 1)""p. 10(第 10 页)"等等。(2)这里的"with"必须去掉,否则本句就没有主语了。本句属于一种"倒装句型"即"状语(介词短语)+被动语态谓语+主语"。(3)英语中标点符号(常见的是逗号和句号)问题:如果投稿美国、加拿大杂志的话,正确的做法是应放在引号内,这是与汉语不一致的地方。(不过英国人的用法与汉语类同,放在引号外。)又如:

Fig. 1 - 3 shows a toy "water rocket $\boxed{.}$"

图 1 - 3 画出了一枚玩具"水火箭"。

This is known as "tuning $\boxed{,}$" which will be discussed in detail in the next section.

这被称为"调谐",我们将在下一节详细地讨论它。

**【改正后的句子】**　**In Part II are provided two choices "Yes" and "No."**

**例 20**

**【汉语原句】**　设 $x_1 = x_{k+1}$,然后走向第二步。

**【英语错句】**　Set $x_1 = x_{k+1}$, then go to the step 2.

**【错误分析】**　(1)"then"之前应该加个"and",或者把逗号改为句号,后面的内容另列一句,这也是读者常犯的一个小错误。(2)定冠词"the"必需去掉

（与上例理由相同）。

【改正后的句子】 **Set $x_1 = x_{k+1}$, and then go to step 2.**

例 21

【汉语原句】 本文提出了一个完善的密码协议，即使信道并不可靠，它也能实现真正的公平性。

【英语错句】 This paper proposed a perfect protocol. It can realize the true fairness even the channels are not reliable.

【错误分析】 （1）"proposed"应该改成一般现在时形式"proposes"。(2)"即使"应该是"even if"而不是仅仅"even"一词。(3)在非特指的不可数名词"true fairness"之前不加定冠词。(4)"信道"在此是泛指，所以在名词复数的情况下不要加定冠词。(5)如果用一个非限制性定语从句来表示第二句的话就能使整个句子紧凑了，否则读起来感到有点松散。

【改正后的句子】 **This paper proposes a perfect protocol，which can（be used to）realize true fairness even if channels are not reliable.**

例 22

【汉语原句】 空气把压力施加于四面八方。

【英语错句】 The air exerts the pressure toward every direction.

【错误分析】 （1）"空气"和"压力"在此均为泛指，所以其前面不该有定冠词。(2)在"direction"前只能用"in"来表示"在……方向上"或"朝……方向"，这属于词汇搭配问题。

【改正后的句子】 **Air exerts pressure in every direction.**

例 23

【汉语原句】 这些窗子均经受不了这么大的力。

【英语错句】 All these windows cannot withstand so large force.

【错误分析】 （1）根据汉语原句可知本句属全否定含义，因此在英语句中不能用"all"与"not"，它一般只表示部分否定"并非都"，由于这里在名词"window"前已有指示代词"these"，所以只能采用"None of these windows"的表达方式。(2)在物理学中，"力"这个概念往往是可数的，例如"一个力可以分解成几个力""两个力可以合成一个力"等等，所以在此应使用"a force"，而由于"so"修饰"large"，所以不定冠词"a"应处于特殊位置上，写成"so large a force"。

【改正后的句子】 **None of these windows can withstand so large a force.**

## 4. 冠词"a"与"an"用法错误

到底用"a"还是用"an",完全取决于其后面单词的第一个"音素"而不是该单词的第一个字母。

### 例 24

【汉语原句】　他们难以在一小时内解决这个问题。

【英语错句】　They are difficult to solve this problem in a hour.

【错误分析】　(1)"They are difficult/easy to do something"是中文式的错句,应该使用不定式复合结构句型来表示,即"It is difficult[easy] for them to do something"。(2)"hour"读音的第一个音素是个双元音,字母"h"是不发音的,所以此句中"hour"前面应该用"an"。

【改正后的句子】　**It is difficult for them to solve this problem in an hour.**

### 例 25

【汉语原句】　本文提出了一种改进的边缘层节点状态预估算法。

【英语错句】　A improved state prediction algorithm of edge layer nodes is proposed in this article.

【错误分析】　(1)"A"应该改为"An"。(2)"algorithm"后应该跟"for"而不是"of"。(3)"(学术)论文"大多数人用"paper"来表示,只有少数人用"article"一词。

【改正后的句子】　**An improved state prediction algorithm for edge layer nodes is proposed in this paper.**

### 例 26

【汉语原句】　我们是否能构造出一个 LFM 信号呢?

【英语错句】　Whether can we construct a LFM signal?

【错误分析】　(1)原文作者由于受到汉语"是否"的影响而写成"whether…"了,实际上"whether"是不能引出疑问句的(这也是读者常犯的错误),这里的"是否"是靠一般疑问句来表达的,即"Can we…?",不过最好写成"Is it possible…?"。(2)句中的"a"应该改成"an"。

【改正后的句子】　**Is it possible for us to construct an LFM signal?**

**例 27**

**【汉语原句】**　该方法把整个因特网分成若干个区域，给每个区域分配了具有唯一长度的固定的区域编码。

**【英语错句】**　The scheme divides the whole Internet into regions, and every region is assigned an unique fixed length regional code.

**【错误分析】**　（1）首先，本句用"and"连接了两个并列的分句，这就显得比较松散，而且两个分句的语态不同，所以最好把"and"后的那个分句改成一个"with 结构"作一种附加说明。（2）"唯一长度"应该作"编码"的后置定语，同时"unique"的读音的第一个音素是辅音[j]，因此不应该用"an"；（3）本句中的"每个区域"强调了个体，所以最好用"each"。

**【改正后的句子】**　**The scheme divides the whole Internet into（several）regions, with each region assigned a fixed regional code of a unique length.**

**例 28**

**【汉语原句】**　当压缩幂次 N 为奇数时，在某些不同条件下，该状态的第一、二分量能呈现出等奇数次幂高次振幅压缩效应。

**【英语错句】**　When squeezing power N is an odd number, under the certain different conditions, the first and second components of state can display equal-odd-power higher-order amplitude squeezing effect.

**【错误分析】**　（1）由于"squeezing power"后面跟有同位语"N"，所以其前面要有定冠词。（2）由于"某些不同条件"并没有特定，所以其前面不必加定冠词。（3）在"state"前应该加定冠词，表示"该状态"；（4）由于"effect"为可数名词单数，因此在其最前面应该加有不定冠词。

**【改正后的句子】**　**When the squeezing power N is an odd number, under certain different conditions, the first and second components of the state can display an equal-odd-power higher-order amplitude squeezing effect.**

**例 29**

**【汉语原句】**　图 2-7 画出了 $x$ 波段波导测试系统。

**【英语错句】**　A $x$-band wave-guide test system is shown by Fig. 2-7.

**【错误分析】**　（1）根据"$x$"的读音，其第一个音素是元音[e]，所以它之前必须用"An"。（2）"by Fig. 2-7"要改成"in Fig. 2-7"，这是英美人的习惯，虽然用主动语态句时可说"Fig. 2-7 shows …"，但被动时一定要用"in"来代替"by"。又如："This paper/book describes …"，变成被动句时只能说"… is described **in** this paper/book"。

【改正后的句子】　**An *x*-band wave-guide test system is shown in Fig. 2 – 7.**

例 30

【汉语原句】　根据自适应 OFDM 系统实时控制的要求,提出了一种基于载波 OFDM 系统信噪比盲估计方法。

【英语错句】　According to the requirement of control in adaptive OFDM system, a SNR blind estimation technique based on virtual carriers is proposed.

【错误分析】　(1) 在名词"requirement"后面英美人一般跟介词"for"。(2) 在"control"前应该加上"the real-time"。(3) 由于"system"是可数名词单数,所以在"adaptive"前应该加上冠词(在此加"an")。(4) 在"SNR"前由于读音的关系应该使用"an"。

【改正后的句子】　**According to the requirement for the real-time control in an adaptive OFDM system, an SNR blind estimation technique based on virtual carriers is proposed.**

# Ⅱ、词汇的选择及用法

例 31

【汉语原句】　在这种情况下,所建议的算法的性能保持鲁棒的。

【英语错句】　In this case, the performance of proposed algorithm keeps robust.

【错误分析】　(1) 在"proposed"之前必须加定冠词。(2) "keeps"是及物动词,可以改成"is kept",这是不少读者常犯的错误,大多英美文章使用连系动词"remain"。

【改正后的句子】　**In this case, the performance of the proposed algorithm remains robust.**

例 32

【汉语原句】　我们详细地分析了选择算法。

【英语错句】　We have analyzed detailedly the algorithm of selection.

【错误分析】　(1) 不存在"detailedly"一词,应该把它改为固定词组"in

detail"。(2)"of"应改为"for",这是"algorithm"所要求的。

**【改正后的句子】** We have analyzed in detail the algorithm for selection.


例 33

**【汉语原句】** 圆形电感器的最大品质因素比方形线圈高出约 20%。

**【英语错句】** The maximum quality factor of the circle inductors is higher about 20% than the square inductors.

**【错误分析】** (1)"circle"应改为形容词"circular",一般来讲,有形容词的就不用名词作定语,类似的有 analytical、mathematical、physical、experimental、calculational 等。(2)"higher"应该放在"than"之前。(3)在"than"后应该加上"that of",否则比较对象就不一致了。

**【改正后的句子】** The maximum quality factor of the circular inductors is 20% higher than that of the square inductors.


例 34

**【汉语原句】** 找到了每个网格单元的热平衡方程,它涉及太阳辐射、地面反射、天空辐射等。

**【英语错句】** The heat balance equation for each grid unit is found. The equation includes solar radiation, ground reflection, sky radiation, and etc.

**【错误分析】** (1)为了使英语句子紧凑,应该把两个句子变成一个句子,也就是把第一个句号变成逗号,然后把"The equation"改成"which"。(2)"涉及"在此应该使用"involves"而不该用"includes"。(3)凡使用了"etc.",则其前面就不能使用"and"一词。当然在此也可以用"and so on"来代替"etc.",因为"and so on"就等效于"etc."的意思。

**【改正后的句子】** The heat balance equation for each grid unit is found, which involves solar radiation, ground reflection, sky radiation, etc.


例 35

**【汉语原句】** 最后,给出了冷却塔表面的温度场,同时对各种相关因素进行了分析。

**【英语错句】** At last, the temperature field on the surface of cooling tower is given, and analysis of various correlative factors is made.

**【错误分析】** (1)本句中一个最常见的错误就是"at last",不少人经常犯这个错误;也有人使用"in the end"。这两个词组的含义是"最终"(=

eventually)，而在这里的"最后"的确切含义是"最后一点"或"最后一项"，所以应该使用"finally"一词，绝大多数英美论文中都是使用这个词，只有极个别使用"lastly"或"last"，这也是正确的。(2)在可数名词"cooling tower"前没有冠词。要记住，一般来说在可数名词单数前一定要有冠词(个别特殊情况例外)，在此既可用"a"，也可用"the"。(3)"作了[进行了]分析"应该使用"make an analysis of …"，这里应该加上不定冠词"an"，这是一个习惯表达法。(4)为了使本句紧凑，把第二个并列分句改成一个"with结构"。"with结构"在现代英语中使用得特别频繁，其构成形式是"with[without]＋名词(或代词)＋分词(或介词短语；形容词；副词；不定式；名词)"；它在句中主要作状语(可处于句首、句中或句尾)。放在句首或句中作状语时最主要表示"如果""当……时候""在……后""因为""虽然"；放在句尾时可表示附加说明、条件、方式等。到底表示什么意思，只能通过试探来确定。它还可放在名词后作定语。例如：

**With the battery terminals** reversed, the result will be very different.

如果把电池端点反过来的话，其结果会是很不相同的。

**With the base** grounded, transistor Q4 is a very high impedance.

当基极接地时，晶体管 Q4 便成为一个非常高的阻抗。

**Without the air** to stop some of the sun's heat, every part of the earth would be burning hot when the sun's rays strike it.

要不是空气挡住太阳的一部分热量，在阳光照射时地球的每一部分都会灼热不堪。

Let us construct, **with the origin** as center, a semicircle of radius R.

让我们以原点为圆心，作一个半径为 R 的半圆。

Standard screws are all right-handed, **with left-hand ones** employed only for special purposes.

标准的螺钉都是右旋的，左旋螺钉只用于特殊的目的。

This current gain shall be measured **with E_p** grounded.

应该在 $E_p$ 接地的情况下来测量这个电流增益。

This is a triangle **with the apex** up.

这是一个顶点朝上的三角形。

【改正后的句子】　**Finally, the temperature field on the surface of a cooling tower is given, with an analysis of various correlative factors made.**

**例 36**

【汉语原句】　本文提出了雷达目标的一种新的识别方法。

【英语错句】　A new kind of recognition method of radar target is presented.

【错误分析】　(1)凡是提到"一种方法""一种语言""一种算法"等等时，这里的"种"字是不该译出来的。(2)在"method"后最常见的是跟"for＋动名词或动作性名词"，也可跟"of＋动名词或动作性名词"，还可跟动词不定式。它后面不能跟"of＋一个具体的东西"，这是属于"中国式英语"表达法。所以应该改成"for the recognition of …"。(3)"radar target"是一个可数名词词组，可以加"a"或用复数形式。

【改正后的句子】　**A new method for the recognition of a radar target（或：for recognizing a radar target）is presented.**

例 37

【汉语原句】　仿真结果表明，本方案具有和其他方案可比的性能。

【英语错句】　The simulation results show that this scheme can obtain comparable performance with other schemes.

【错误分析】　(1)去掉句首的"The"，因为这不是特指。(2)从真实的含义上看，应该把"obtain"改成"lead to"或"result in"。(3)"comparable performance with"应改成"a performance comparable to that of"。

【改正后的句子】　**Simulation results show that this scheme can lead to a performance comparable to that of other schemes.**

例 38

【汉语原句】　对构件的承载能力作了具体的分析。

【英语错句】　The detailed analysis of the ability of bearing loads of the components is given.

【错误分析】　(1)"作一分析"应该表示为"make an analysis"，这属于名词"analysis"与动词"make"搭配使用。又如"作一研究"为"make a study"，"作一比较"为"make a comparison"；但是"作一说明"则要用"give an explanation"，"作一叙述"为"give a description"，"作一介绍"为"give an introduction"。所以这里的"given"应改成"made"。(2)"承载"应该是"carry loads"而不是"bear loads"。(3)在"能力(ability；capacity)""趋势(tendency)""愿望(desire)""未能(failure)""意图(intention)"等词后一般应使用由"of"引出的"类似性动词不定式复合结构"作后置定语，其模式为"the ability **of A to do B**"。例如：

Energy is defined as the <u>ability</u> **of a body to do work.**

能量被定义为**物体做功的**<u>能力</u>。

Gravity is the <u>tendency</u> **of all objects to attract，and be attracted by，each other.**

引力就是**所有物体相互吸引的**<u>趋势</u>。

The <u>desire</u> **of man to control nature's forces successfully** has been the catalyst for progress throughout history.

**人类想要成功地控制大自然的各种力的**<u>愿望</u>，一直是推动整个历史进程的催化剂。

The deviations from the expected periodicity in Mendeleev's list were due to the <u>failure</u> **of contemporary chemistry to have discovered some of the elements existing in nature.**

那时之所以与门捷列夫周期表中所预言的周期性有出入，是由于**当时化学界**<u>未能</u>**发现存在于自然界的某些元素**之故。

The apparent <u>intention</u> **of the gas to shrink to zero volume at absolute zero** is naturally never fulfilled.

**气体在绝对零度时体积缩小到零的**<u>明显</u>**意图**，自然是永远不可能实现的。

【改正后的句子】　**A detailed analysis of the ability of the components to carry loads is made.**

### 例 39

【汉语原句】　其余四个参数可以通过解一组特殊的线性方程而计算出来。

【英语错句】　The rest four parameters can be calculated by solving a set of special linear equations.

【错误分析】　(1)"其余的"不能用"rest"，这是不少人经常用错的，它作名词时要用"the rest"，意为"其余""其余的人""剩余部分"。因为"rest"作定语时在物理学中的含义是"静止的"，而在生活上的含义是"休息的"，所以这里应该用"remaining"。(2)形容词"special"一定要放在"set"前而不能放在"linear equations"前，这同样是不少人经常出错的地方，这是英美写作的一个习惯用法。常见的是形容词应该放在 kind，type，set，sort，form，pair，series 等前面，如："一种新的半导体材料"为"a **new kind** of semiconductor material"，"一种新型的电动机"为"a **new type** of electric motor"，"一套完整的工具"为"a **complete set** of tools"，"一双新鞋"为"a **new pair** of shoes"。

【改正后的句子】　**The remaining four parameters can be calculated by solving a special set of linear equations.**

**例 40**

【汉语原句】　共有 $M$ 个多边形，其顶点数均为 $N$。

【英语错句】　There are $M$ polygons altogether, whose vertex number is all $N$.

【错误分析】　句中的那个定语从句是一个典型的"中国式英语"句子，其中有两个错误：(1)"顶点数"并不是"vertex number"（这是"顶点的号码"之意），而应该是"the number of vertexes"，同样"student number"意为"学号"，而"the number of students"才是"学生数"的意思。(2)在这里绝不能用"all"来表示"均"的含义，而是应该使用"each"来表示。

【改正后的句子】　**There are $M$ polygons altogether, each of which has $N$ vertexes.**（虽然从语法角度来讲，对于那个定语从句我们也可以把它表示成"the number of vertexes of each of which is $N$"，但是这一表示法显得太繁杂了。）

**例 41**

【汉语原句】　本文对提高打印速度和印字质量提供了一种机辅设计手段。

【英语错句】　This paper provided a kind of means of computer aid design for progressing the printing speed and the printing quality.

【错误分析】　(1)一般在使用"本文讨论了……"这种句型时，不用过去时而应该使用一般现在时，所以"provided"是不妥的，这里最好使用"presents"这一动词。(2)"一种机辅设计手段"事实上就是"a CAD method"，在此句中不能使用"kind"一词。(3)"progressing"这个词用错了，因为"progress"当动词用时为不及物动词，要表达"提高速度"应使用动词"increase"，而要表示"提高质量"应使用"improve"，这属于词汇搭配问题。

【改正后的句子】　**This paper presents a CAD method for increasing the printing speed and improving the printing quality.**

**例 42**

【汉语原句】　若自偏电路设计不当，LC 振荡器会出现间歇振荡。

【英语错句】　If the self-bias circuit does not design good, the LC oscillator will take place chopping oscillations.

【错误分析】　(1)本句属于典型的"中文式的英语句"。在 if 引导的条件状语从句中，"电路"自己是不会"设计"的，所以应该使用被动语态或别的句型。(2)"设计"在句中为动词，则修饰它的应该是副词 well 而不能用"good"。

（3）在主句中，"take place（发生）"为不及物性的短语动词，所以不可能有宾语。

**【改正后的句子】** If the self-bias circuit is not designed well（或：properly），（或：… is of poor design，）chopping oscillations will take place in the LC oscillator.

### 例 43

**【汉语原句】**　现在让我们把所提出的算法的性能与文献[1][3]中的算法作一比较。

**【英语错句】** Now let's compare the performance of the proposed algorithm with the algorithms in literatures [1] [3].

**【错误分析】**　（1）在科技文中一般要把"let's"写成"let us"，这是写惯了普通英语后所经常出现的问题。（2）比较的对象不一致，所以应该在"with"后面加上"that of"，表示"性能"与"性能"进行比较。（3）名词"literature（文献）"与"information（信息）"一样永远属于不可数名词，因此不能有复数形式的。

**【改正后的句子】**　Now let us compare the performance of the proposed algorithm with that of the algorithms in literature [1][3].（更好的版本为：Now let us make a comparison in performance between the proposed algorithm and those in literature [1][3].）

### 例 44

**【汉语原句】**　两小时后该机器会自动地再次启动。

**【英语错句】**　The machine will automatically start to operate again two hours later.

**【错误分析】**　本句的错误之处在于"later"一词。该词可以用于将来时态的句子中，但其前面不能带有表示时间长短的数量状语（在过去时的句子中可以这样使用），这时应使用介词短语"in two hours"来表示"两小时后"。一般来说，在将来时态的句子中，若句中的动词属于"终止性动词（如 leave，start，begin，come，see 等）"或表示状态的"系表结构"时，"in"表示"在……之后"之意。例如：

They will leave for Shanghai **in** a few days.

他们将在几天后前往上海。

Our president will be back **in** two days.

我们的校长将在两天后回来。

若句中的动词为"延续性动词（如 learn，study，draw，design，compute

等)", 则"in"表示"在……之内"之意。例如:

Mr. Wang can draw a beautiful horse **in** five minutes.

王先生能在五分钟之内画出一匹漂亮的马。

The research institute will design a new type of computer **in** two months.

该研究所将在两个月内设计出一种新型的计算机。

【改正后的句子】 **The machine will automatically start to operate again in two hours.**

例 45

【汉语原句】 作者在此要感谢 W. Smith 教授所给予的帮助。

【英语错句】 The author here thanks to professor W. Smith for help.

【错误分析】 (1)"thank"作动词时为及物的,所以在其后面不该加介词"to";当它作名词时,则要用复数形式"thanks"后跟介词"to"。(2)"主席""教授""医生""博士"等作头衔时的第一个字母一定要大写,所以这里应该写成"Professor W. Smith"。(3)"here"在此是多余的,应该去掉。(4)一般在这里的"help"前应该加上"his"才符合英美文章的习惯。常见的表示感谢的词汇有 thank, thankful, thanks, acknowledge, acknowledgement, gratitude, grateful, appreciate, appreciation, indebted, indebtedness, recognition 等。表示感谢的句型很多,常见的有以下一些:

I'd like to thank…

我要感谢……

The author wishes to express his sincere thanks to…

作者要对……表示由衷的感谢。

I'd like to acknowledge…

我要感谢……

The author wishes to make the acknowledgement…

作者要感谢……

Grateful acknowledgement is made to…

作者对……深表感谢。

The author is in acknowledgement of…

作者要感谢……

I would like to express my gratitude to…

我要感谢……

Special gratitude is owed to…

作者要特别感谢……

A special gratitude is expressed to…

作者要特别感谢……

It is a pleasure to record my gratitude to…

我愿感谢……

I'm grateful to…

我要感谢……

I gratefully appreciate…

我对……极为感激。

I would like to express my appreciation of…

我要感谢……

I'm greatly indebted to…

我对……深表感谢。

I'm in indebtedness to…

我要感谢……

We express our indebtedness to…

我们要感谢……

I wish to thank…

我要感谢……

Thanks are owed to…

作者要感谢……

Thanks are due to …

作者要感谢……

Thanks go to…

作者要感谢……

Thanks must be given to…

作者要感谢……

Thanks must be extended to…

作者要感谢……

Special thanks go to…

作者要特别感谢……

I offer my thanks to…

我要感谢……

A word of thanks is given to…

作者要感谢……

Particular recognition is due to…

作者要特别感谢……

We wish to give special thanks to…

我们要特别感谢……

His sincere thanks are offered to…

作者对…表示衷心的感谢。

We are thankful to…

我们要感谢……

作者若要对几方面的人士分别表示感谢的话，为了避免语言上的死板、乏味，往往要使用上面提到的不同句型来表示"感谢"这一含义。

**【改正后的句子】** **The author would like to express his thanks to Professor W. Smith for his help.（或：The author would like to thank Professor W. Smith for his help.）**

**例 46**

**【汉语原句】** 由于这些理由，希望计算机能把文本与图像分离开来。

**【英语错句】** Since these reasons, it has become desirable for a computer to separate the text from images.

**【错误分析】** 句中"Since"应该改成"For"（"reason"前的搭配介词是"for"），因为"since"作为介词时它的词义为"自从"，没有"由于"之意，只有当它用作从属连词时才具有"自从"和"既然，因为"两个词义。有些读者在写作时常犯这一错误，由于他们只记住了"since"有"因为"这一词义而不清楚它的词性。同样有的读者使用"because＋名词"的错误表示法。当然本句也可以改为"That is why it has become desirable…"，这样语气就被加强了。

**【改正后的句子】** **For these reasons, it has become desirable for a computer to separate the text from images.**

**例 47**

**【汉语原句】** 这种方法对工程制图的种类没有多少限制条件。

**【英语错句】** This method has little limitations to the kind of engineering drawings.

**【错误分析】** (1)"limitation"在此是一个可数名词，可以用复数形式，因此句中用"little"来修饰它是错误的，应该改为"few"。(2)"limitations"后的搭配介词通常是"on"而不是"to"(在名词"limit"之后则通常使用介词"to"的)。(3) 最好把动词"has"改为"imposes"或"puts"似乎就更合原意。(4) 把"kind"改为复数形式更合句意。

**【改正后的句子】** **This method imposes few limitations on the kinds of engineering drawings.**

### 例 48

**【汉语原句】** 卡尔曼滤波器(UKF)具有易实现、较高的估计精度和中等的计算量等优点。

**【英语错句】** The advantages of recently developed UKF are significant with its ease to implement, better accuracy, some order of computational complexity.

**【错误分析】** 整个句子安排不对,重点应该说明其优点是什么,所以本句要进行大的改动。(1)"significant"应该放在"advantages"之前。(2) 在"recently"之前应该加冠词"the",因为"UKF"是一个可数名词单数。(3) 去掉"with",把"its ease to implement"改成"its easy implementation"。(4)"some order of"改成"moderate"。(5) 在最后两个优点之间应该加上"and"。

**【改正后的句子】** **The significant advantages of the recently developed UKF are its easy implementation, better accuracy, and moderate computational complexity.**

### 例 49

**【汉语原句】** 对 PRR,ERR 和 MBRR 三种方案的性能作了比较。

**【英语错句】** Comparison is done between the performance of PRR, ERR and MBRR schemes.

**【错误分析】** (1)"Comparison"一般应写成"A comparison"。(2)"作比较"的惯用表达式是"make a comparison"而不能用动词"do"。(3)"done between the performance of"应改成地道的表达法"made in performance between…and …"。

**【改正后的句子】** **A comparison is made in performance between PRR, ERR and MBRR schemes.**(这里可以把"between"改成"among"。)

### 例 50

**【汉语原句】** 用这种方法所需的迭代次数为现有方法的 10%。

【英语错句】 The number of iterations by this method is 10% of the relevant methods.

【错误分析】 (1)在"by this method"前应加上"required",表示"所需的"之意。(2)"10%"后面的"of"一般是可以省去的,但要加上"that required by …"。(3)表示"现有的"常用"available"一词,且后置,所以把"relevant methods"改成"methods available"。

【改正后的句子】 **The number of iterations required by this method is 10% (of) that required by the methods available.**

例 51

【汉语原句】 针对杂波功率随距离缓变的非均匀环境,提出了一种新的加权最大似然估计算法。

【英语错句】 For the clutter power slowly changed environment,a novel weighting maximum likelihood estimation (WMLE) algorithm is proposed.

【错误分析】 "the clutter power slowly changed environment"是中国式的英语表示法,把词汇按汉语次序堆起来的,而且"非均匀"这一含义并没有表示出来,应该把它改为"the non-uniform environment in which[where]the clutter power changes slowly with distance"或更好地译成"the non-uniform environment in which[where]the variation of clutter power with distance is slow"。

【改正后的句子】 **For the non-uniform environment where/in which the variation of clutter power with distance is slow,a novel weighting maximum likelihood estimation (WMLE) algorithm is proposed.**

例 52

【汉语原句】 这不仅有利于区分开每个信号项,而且有利于减少截项。

【英语错句】 This is not only benefit to distinguish each signal term and but also to reduce cross terms.

【错误分析】 (1)"not only…but also"属于并列连接词,所以两部分在句中所处的位置应该相同,因而在句中应该把"not only"放在"to"之前。(2)"有利于"应该使用形容词短语"beneficial to",这个"to"是介词而不是动词不定式的标志,所以其后应该跟动名词而不能跟动词不定式。同样的情况有"helpful to","conducive to"等,所以学习时一定要注意"to"到底是介词还是动词不定式的标记。(3)句中的"and"是多余的。(4)"区分开每个信号项"可以由两种方法来表示，一种是使用"distinguish signal terms from one another（也可以用 each other）"，另一种是使用"distinguish between signal terms"。

【改正后的句子】　**This is beneficial not only to distinguishing signal terms from one another，but also to reducing cross terms.**

例 53

【汉语原句】　首先将集总网络的频域导纳参数写成多个有理分式之和的形式，通过傅里叶变换技术将其变换到时域。

【英语错句】　At first，the admittance parameters in frequency domain of the lumped networks are expressed as a summation of several rational fractions，which are converted into the time domain by means of the inverse Fourier transform technique.

【错误分析】　（1）"At first"意为"起初，开始的时候"，应该把它改为"First"。（2）在"frequency domain"前一定要加"the"。（3）"which are converted into the time domain"应改为"which is converted into that in the time domain"，因为它指的是"summation"。（4）去掉"means of"。

【改正后的句子】　**First，the admittance parameters in the frequency domain of the lumped networks are expressed as a summation of several rational fractions，which is converted into that in the time domain by the inverse Fourier transform technique.**

例 54

【汉语原句】　这个方案难以抵制节点捕获、DoS 和信息重放等攻击。

【英语错句】　This scheme is difficult to resist node capture，query replay and DoS attacks.

【错误分析】　（1）"This scheme"应改为形式主语"It"。（2）在"to resist"前加"for this scheme"，构成不定式复合结构作主语。（3）在"to resist"后加"attacks such as"。

【改正后的句子】　**It is difficult for this scheme to resist attacks such as node capture，query replay and DoS.**

例 55

【汉语原句】　为了讨论方便起见，设映射矢量为 $Z$ 轴，因而相应的 $Y$ 轴就是侧向轴。

【英语错句】 In order to discuss conveniently, let the projection vector to be the Z-axis, therefore, the corresponding Y-axis is the cross-range axis.

【错误分析】 (1)"为了讨论方便起见",在错句中属于中文式的英语表示法,因为在及物动词"discuss"后面没有宾语,所以这里应该使用介词短语"for the convenience of discussion"。(2)"to be"应该改成"be",因为"let"要求的宾语补足语是不带"to"的不定式。(3)在"therefore"前应该加连接词"and",这时一般把"therefore"后的逗号去掉;或者也可以把第二个逗号改为分号或句号(改成句号后,要把"therefore"的第一个字母大写)。不过这里用"so that"引出结果状语从句似乎更好些。

【改正后的句子】 **For the convenience of discussion, let the projection vector be the Z-axis so that the corresponding Y-axis is the cross-range axis.**

例 56

【汉语原句】 所获得的耦合参数分成两大类。

【英语错句】 The resulted parameters are divided into two big classes.

【错误分析】 (1)由于"result"作动词时是不及物的,所以不能用它的过去分词作定语,而只能用现在分词"resulting"或形容词"resultant"。(2)"两大类"中的"大"在英语中通常可用 broad、general、main、major 等来表示,一般不用"big"。(在这里用"fall"来代替"are divided"似乎更理想。)

【改正后的句子】 **The resulting parameters fall into two broad classes.**

例 57

【汉语原句】 这台新设计的仪器质量很好。

【英语错句】 This new designed instrument is in good quality.

【错语分析】 (1)"新"在此是修饰"设计"的,由于"designed"属过去分词作前置定语,相当于形容词的性质,所以修饰它的词必须是一个副词,因此应该把"new"改成"newly"。(2)表示"质量好"可有两种比较好的方法,一种是用"be good in quality",另一种是"be of good quality";当然也可以用"The quality of … is good"来表示,但不如前两种表示法好。

【改正后的句子】 **This newly designed instrument is good in quality.**(或:**This newly designed instrument is of good quality.**)

例 58

【汉语原句】 这个问题有待解决。

**【英语错句】** This problem waits to solve.

**【错误分析】** 这句是一个典型的中国式英语表示法。(1) 表示“有待于”概念时应使用半助动词“remain”后跟动词不定式。(2) 本句的主语“这个问题”是“解决”的逻辑对象，所以应该使用被动的形式“to be solved”。

**【改正后的句子】** **This problem remains to be solved.**

**例 59**

**【汉语原句】** 该系统能很好地阻止多路径传播，并能在系统噪声比为 −20～30 dB时，定位误差小于 60 cm。

**【英语错句】** The system is good to resist multi-path propagation and can achieve position location error under 60 cm with SNR varying −20 dB to 30 dB.

**【错误分析】** (1)“能很好地阻止”应该表示成“can prevent[stop]in a good way”。(2)“position location error”应改为“a location error of”。(3)“with”应改为“for”。(4)“小于”一般表示成“less than”。(5) 在“SNR”前应该加“an”。

**【改正后的句子】** **The system can prevent multi-path propagation in a good way and achieve a location error of less than 60 cm for an SNR varying from −20 dB to 30 dB.**

**例 60**

**【汉语原句】** 与传统的最优合并方案不同，这种方案不需要信道参数。

**【英语错句】** Different from conventional optimum combining scheme, channel parameters are not needed in this scheme.

**【错误分析】** (1) 由于“Different from …”的逻辑主语应该是“this scheme”，所以本句后半部分应该写成“this scheme does not need …”。(2) 在“from”后应该加冠词“the”，或如果传统方案有多个的话则只要把“scheme”改成复数形式而不必加冠词了。

**【改正后的句子】** **Different from conventional optimum combining schemes, this scheme does not need (any) channel parameters.**

**例 61**

**【汉语原句】** 这种新的罗盘精度高、成本低、携带方便。

**【英语错句】** This kind of new compass is accurate, low in cost and taken easily.

【错误分析】 （1）"new"应该放在"kind"之前。（2）"is"应该改成"has"而其后要用名词作宾语，即添加"the features"，然后用"of"引出"features"的同位语，即"high accuracy，low cost and great portability"。

【改正后的句子】 **This new kind of compass has the features of high accuracy，low cost and great portability.**

**例 62**

【汉语原句】 一种新的宽带目标识别雷达杂波抑制方法

【英语错句】 A new kind of method for clutter suppression of wideband target recognition radar

【错误分析】 这是一篇论文的标题。（1）根据我国论文标题的写法，开头的冠词"A"要省去。（2）方法"method"不能用"a kind of …"来修饰。（3）"for clutter suppression of"应改成"for suppression of the clutter …"。（4）由于"……雷达"是个具体的东西，所以其前面要有冠词。

【改正后的句子】 **New method for suppression of the clutter of the wideband target recognition radar.**

**例 63**

【汉语原句】 本文提出了一种新的基于概率的流量均衡算法，并分析了它在有故障的直连网络中的性能。

【英语错句】 A new kind of probability-based traffic balance algorithm is presented in the paper，and its performance in faulted direct networks is analyzed.

【错误分析】 （1）本句的主要错误与上句相同，不能用"kind of"，这是中国人在科技写作时经常犯的一个错误，主要是受了汉语的影响而逐字翻译的结果。（2）最好把"and"之后的东西用一个"with 结构"来表示，使句子显得紧凑。（3）"有故障的"应该使用"faulty"或"defective"。（4）在英文文摘的第一句中，"in the paper"一般是不必写出来的。

【改正后的句子】 **A new probability-based traffic balance algorithm is presented，with its performance in faulty direct networks analyzed.**

**例 64**

【汉语原句】 图 5 画出了输入电阻随 B 变化的曲线。

【英语原句】 Fig.5 draws the curve of input resistance varied with B.

**【错误分析】**　（1）汉语原句中的"画出了"译成"draws"是典型的中国式英语表示法，而最常见的正确单词是"shows"，也可用"illustrates"来表示。（2）"M 随 N 的变化"应该使用名词与介词的搭配模式，"the variation of M with N"来表示。即使退一步说，在这里也不能使用"varied"，因为"vary"属于不及物动词。

**【改正后的句子】**　**Fig.5 shows the curve of the variation of input resistance with B（或：⋯the curve of input resistance against［versus］B）.**

**例 65**

**【汉语原句】**　这一定义具有重要的意义。

**【英语错句】**　This definition has an important meaning.

**【错误分析】**　（1）当表示具有一个抽象的东西时，往往使用"be of"来表示"具有"之意。（2）在"重要的意义"的表达上，由于写作者受到了汉语的影响采用了逐字对译的错误做法而造成了表达上的错误。英语中"意义"是用"significance"来表示的，这时我们汉语的"重要的"在英语中在此只能用"great"来表示，而不能使用"important significance"，这属于两国语言表达上的不同，望读者阅读原文时要特别留意并多加记忆。对于汉英不同之处我们记得越多，写出的句子就越地道。又如汉语说"太阳离地球的距离是很远的"，应该写成"The distance of the sun from the earth is very great"。汉语中可以说"距离是远的"，而英语中却不能说"The distance is far"。因为"距离"代表了一个数值，它没有远近之分，只有大小、长短之分。

**【改正后的句子】**　**This definition is of great significance.**

**例 66**

**【汉语原句】**　理论分析与实验数据相吻合，从而解决了长期存在的 1/E 和 E 模型之争。

**【英语错句】**　The theory analysis fits well the experiment data, so the long-existed dispute between 1/E and E models is settled.

**【错误分析】**　（1）由于"理论"在此是作定语的，所以不能用名词"theory"来修饰，而应该改成形容词"theoretical"。（2）同样，"experiment"应该改写成"experimental"。（3）"fit"当及物动词时意为"适合于"，而要表示"与……相吻合"应该使用"fit into"，不过多数英美人喜欢使用"agree with"或"be in agreement with"来表示。（4）"exist"是一个不及物动词，所以只能用其现在分词作定语，也就是应该把"long-existed"改成"long-existing"。（5）如果把逗号

以后的东西改成分词短语作结果状语，可以使句子结构精炼而紧凑，即写成
"thus settling the long-existing dispute between 1/E and E models"。

【改正后的句子】 **The theoretical analysis agrees well with the experimental data，thus settling the long-existing dispute between 1/E and E models.**

### 例 67

【汉语原句】 最后指出了上面证明中的一个问题并给出了两种解决方法。

【英语错句】 A problem in above proof is proposed finally with two methods to solve the problem.

【错误分析】 （1）一般应该把"finally"放在句首。（2）在"above proof"之前应该有一个定冠词。（3）这里的"指出"应该用"point out"来表示而不是"propose"（它意为"建议性提出"）。（4）这里的"with 短语"放在被动语态谓语之后成了修饰谓语动词的方式状语，应该使用"with 结构"来表示附加内容。（5）"解决方法"英美人经常使用"solution"一词来表示。

【改正后的句子】 **Finally，a problem in the above proof is pointed out，with its two solutions given.**

### 例 68

【汉语原句】 功等于力乘以距离。

【英语错句】 Work equals to force multiplying distance.

【错误分析】 （1）"equal"这个词既可以是及物动词，也可以是形容词，若把它用作及物动词，则应把上述句中的"to"去掉，若把它用作形容词，则应把"equals"改成"is equal"。（2）这里"A 乘以 B"要用过去分词短语来表示成"A multiplied by B"（当然也可以用"A times B"或"the product of A and B"来表示）。同样"A 除以 B"应表示成"A divided by B"（个别时候也可用"A over B"）。

【改正后的句子】 **Work equals［is equal to］force multiplied by distance.**

### 例 69

【汉语原句】 我们要感谢评阅人对我们的稿子所做的一切。

【英语错句】 We want to thank the referees for all that you do for our manuscript.

【错误分析】 （1）表示"要感谢"的句型有好多（在前面已介绍过），为了客

气，一般应该把"want"（不少中国人喜欢用这个动词）改成"would like"或"wish"。（2）"you"必须改成"they"，因为这里是复数"referees"。（3）应该把"do"改成"have done"，说明已经做了的事。

**【改正后的句子】　We would like to thank the referees for all that they have done for our manuscript.**

**例 70**

**【汉语原句】**　在这种情况下也会存在由于为数有限的模式而引起的估计误差。

**【英语错句】**　In this case，there also exists estimation error resulted by the finite number of patterns.

**【错误分析】**　（1）"error"是可数名词单数，所以应该带有一个冠词（这里最好加"an"）。（2）由于"resulted"是不及物动词的过去分词，所以不能用作定语的，可以把它改成"caused"，或者把"resulted by"改成"resulting from"（主动的形式、被动的含意）。（3）"the finite number of"应改成"a finite number of"，又如："a number of（一些，几个）""a small number of（少量的）""a large number of（大量的，许多）"。

**【改正后的句子】　In this case，there also exists an estimation error resulting from a finite number of patterns.**

**例 71**

**【汉语原句】**　这种算法与其他 AC 算法相比在带宽利用和网络收益上有明显的优点。

**【英语错句】**　This algorithm has notable advantages both in bandwidth utilization and network revenue comparing with other AC algorithms.

**【错误分析】**　（1）由于词汇搭配"advantage over…"表示"与……相比的优点"，"over"就具有"相比"之意，所以应该把"comparing with other AC algorithms"改成"over other AC algorithms"并放在"notable advantages"之后。（2）"both…and…"是一个并列连接词，它连接的两个成分在语法上应该是一样的，所以要把"both"放在"in"之后，这一错误也是经常出现的。

**【改正后的句子】　This algorithm has notable advantages over other AC algorithms in both bandwidth utilization and network revenue.（或：This algorithm is notably superior to other AC algorithms in both bandwidth utilization and network revenue.）**

**例 72**

**【汉语原句】**　动量因子对 $\Delta W(n-1)$ 的影响与阶跃尺寸对 $\Delta W(n)$ 的影响是相同的。

**【英语错句】**　The effect of the momentum factor to $\Delta W(n-1)$ is same as the step size to $\Delta W(n)$.

**【错误分析】**　(1) 注意词汇搭配模式"the effect of A on/upon B"，所以主语短语后的"to"要改为"on/upon"。(2) 在"same"之前要加定冠词"the"，这是固定用法。(3) 在"as"后要加上"that of"；这样比较对象就一致了，"that"代替了"the effect"。(4) 句中第二个"to"也要改为"on/upon"。

**【改正后的句子】**　**The effect of the momentum factor on $\Delta W(n-1)$ is the same as that of the step size on $\Delta W(n)$.**

**例 73**

**【汉语原句】**　对陈等人提出的计算 Tanner 图中最短圈数量的算法加以改进。

**【英语错句】**　The algorithm to count shortest cycles of Tanner graphs proposed by Chen etc. is improved.

**【错误分析】**　(1) 在"algorithm"后面搭配的是"for"，所以"to count"应该改为"for counting the"。(2) 在"人名＋名词"前要加定冠词，所以在"Tanner"前加上"the"。(3) "etc."是用于事物的，对于人应该使用"et al."。

**【改正后的句子】**　**The algorithm for counting the shortest cycles of the Tanner graphs proposed by Chen et al. is improved.**

**例 74**

**【汉语原句】**　结果表明，这两种方法其实质是一致的，只不过各有其应用的条件。

**【英语错句】**　The results indicate that this two methods are consistent, they have only respective applied condition.

**【错误分析】**　(1) 在逗号之前的宾语从句中的错误："这两种方法"的"这"后面指的是复数名词，所以应该用"these"而不能用"this"，这是受了汉语"这"字的影响而发生的错误。(2) 在句中"其实质"并没有表示出来，可以用"in essence"或"in nature"来表示。(3) 在逗号后的错误：两个并列句之间一般应加并列连接词，在此应该加"but"。(4) "各自的应用条件"可以写成"their own specific application conditions"。逗号后若用"with 短语"则显得更紧凑，即写成："with the exception that they have their own specific application conditions"。

【改正后的句子】 The results indicate that these two methods are consistent in essence, with the exception that they have their own specific application conditions.

例 75

【汉语原句】 设计出了一种新的算法，它能自动地维持有用的副本的生存，从而大大地减少了网内的分组副本数。

【英语错句】 A kind of new algorithm has been designed to be able to keep the life of useful copies. Thus the number of packet copies in a network will be decreased obviously.

【错误分析】 (1)"kind of"应该去掉。(2)动词不定式处于被动语态谓语后是不能作主语的定语的，所以应该把"to be able to…"改成定语从句"which will be able to…"的形式。(3)本句的英语文本用了两个独立的句子就显得松散，应该把第二部分用一个分词短语来表示结果，即"thus greatly reducing the number of packet copies in a network"。(4)"life"应换成"survival"一词，表示"生存"。(5)"keep"改成"maintain"为好，表示"维持"。

【改正后的句子】 A new algorithm has been designed which will be able to maintain the survival of useful copies, thus greatly reducing the number of packet copies in a network.

例 76

【汉语原句】 这台仪器的价格很昂贵。

【英语错句】 The price for this instrument is expensive.

【错误分析】 (1)"the price"后跟"for"的意思是"为……付出代价"，所以应该把"for"改成"of"，这时它才表示"价格"。(2)"expensive"和"cheap"这种形容词只能修饰具体的物品，而不能修饰"价格"，这主要是受了汉语的影响，"价格(price)"这个词只能由"high"或"low"来修饰。

【改正后的句子】 The price of this instrument is very high.

例 77

【汉语原句】 这个设备占地方太多。

【英语错句】 This device takes too many places.

【错误分析】　（1）这里的"地方"是不能用"place"一词的，这也是受了汉语的影响，应该用"space""room"。（2）"many"应改为"much"，因为这是不可数的含义。（3）表示"占据"一般使用"take up"（当然"take"一词也有此意的），不过在科技文中常用"occupy"一词。

【改正后的句子】　**This device occupies too much space[room].**

例 78

【汉语原句】　在这种情况下，必须尽量利用信号的能量。

【英语错句】　In this case it is necessary to use the energy of signals as many as possible.

【错误分析】　"as…as possible"这一表达形式应该套在"energy"上面，而由于"energy"在此为不可数名词，所以应该使用形容词"much"。

【改正后的句子】　**In this case it is necessary to use as much energy of signals as possible.**

例 79

【汉语原句】　甚至在这种情况下，这些协议也能为有限数量用户提供可接受的话音服务质量。

【英语错句】　Even in this case these protocols can provide limited amount of users acceptable quality of voice service.

【错误分析】　（1）"provide"这个动词使用的搭配模式应该是"provide sb. with sth."或"provide sth. for sb."，所以在"users"之后应加上介词"with"。（2）"有限数量用户"表明这个"数量"是说明可数名词"用户"的，因此在这里一般不能用"amount"而应该用"number"。（3）在"limited"之前应使用不定冠词"a"，因为"a…number of"属于一种固定表示法，起形容词含义的作用。（4）在"acceptable"之前应该使用定冠词"the"。

【改正后的句子】　**Even in this case these protocols can provide a limited number of users with the acceptable quality of voice service.**

例 80

【汉语原句】　均衡器对均衡性能的影响进行了详细的分析。

【英语错句】　The effect of equalizer parameters for equalization performance is analyzed detailedly.

【错误分析】　（1）本句主要的错误在于名词"effect"与介词的搭配搞错了，

其搭配模式必须是"the effect of A on/upon B",意为"A 对 B 的影响",所以本句中的"for"必须改成"on"或"upon"。(2)"详细地"一词的译文应该使用常用的固定词组"in detail",虽然词典上有"detailedly"这个副词,可是实际上英美科技人员是不用的。又如"困难地"应使用"with difficulty",美国外教说"我们从来不用'difficultly'这个副词的"。所以希望读者在阅读英美原文时多观察、多记忆。

**【改正后的句子】 The effect of equalizer parameters on equalization performance is analyzed in detail.**(更好的版本为:**A detailed analysis is made of the effect of equalizer parameters on equalization performance.** 为了避免句子"头重脚轻"现象,这里使用了分隔句型)

### 例 81
**【汉语原句】** 将一类非线性椭圆方程组的求解问题转化为求一给定泛函的临界点问题。

**【英语错句】** The solving problem of a class of system of nonlinear elliptic equations is transformed into a problem of critical points of given functional.

**【错误分析】** (1)"solving problem"属于中文式的英语表达法,应该把它改成"the solution of … problem"。(2)在"critical"前应该加上动名词"finding"。(3)在"given"前应该加上不定冠词"a"。

**【改正后的句子】 The solution of a class of system of nonlinear elliptic equations is transformed into a problem of finding critical points of a given functional.**

### 例 82
**【汉语原句】** 最后获得了鼓舞人心的结果,同时给出了电路实例来阐明该模型。

**【英语错句】** At last, encouraged results have been obtained, circuit examples are given to demonstrate the model.

**【错误分析】** (1)"At last",应该改成"Finally",在前面的例句中已有说明。(2)"encouraged"应该使用现在分词"encouraging",有些读者经常搞错,其简单的判别法是:修饰人时用"…ed"形式;修饰事物时用"…ing"形式。例如:

That phenomenon is very **interesting**.

那个现象是很有趣的。

They are **interested** in that phenomenon.

他们对那个现象感兴趣。

The experimental results are **encouraging**.

实验的结果令人鼓舞。

Those scientists were greatly **encouraged** by the experimental results.

那些科学家为实验结果受到极大的鼓舞。

（3）本句的后半部分应该用一个"with 结构"使句子紧凑而精炼，即："with circuit examples given to demonstrate the model"。

【改正后的句子】　**Finally，encouraging results have been obtained，with circuit examples given to demonstrate the model.**

**例 83**

【汉语原句】　首先构造了一个综合性能函数，然后根据此函数，把各种因素放在一起考虑来建立优化模型。

【英语错句】　At first，a integrated performance function is formulated. According to the integrated performance function，all factors are considered together to establish the optimal model.

【错误分析】　（1）"At first"的含意是"起初"，这里应该用副词"First"（不少读者经常搞错），不过由于改动了后面的东西，因而这个词不必表示出来。（2）句中的不定冠词应该改成"an"。为了使句子紧凑，把原来的第二个句子变成一个非限制性定语从句，即"according to which …"或"on the basis of which …"。（3）"together"这个副词是多余的。如果把"consider"改成"take into consideration[account]"读起来似乎更好一些。

【改正后的句子】　**An integrated performance function is formulated，on the basis of which all factors are taken into account to establish the optimal model.**

**例 84**

【汉语原句】　我们拟合出多普勒频率随距离变化的函数。

【英语错句】　We fit the function of the Doppler frequency rate with range.

【错误分析】　（1）对于"拟合出"应该用"have established [derived]"来表示。（2）"A 随 B 变化"的表示方法有三种：① "variation of A with B"；② "A against B"；③ "A vs B"。

**【改正后的句子】**　We have established the function of the variation of（the）Doppler frequency rate with（the）range.

### 例 85

**【汉语原句】**　同时，本文导出了所有这些因素对系统性能影响的定量表达式，在此基础上提出了一种改进型的系统。

**【英语错句】**　Simultaneously, this paper derived quantitative expression of all these factors' effect on the performance of the system. On the basis of it, the paper presented a kind of improved system.

**【错误分析】**　（1）本句属于一篇文摘中的句子，所以不该用过去时，而应该使用一般现在时。（2）这里表示"同时"一般使用"also"一词，因为它并不表示时间含义，所以一般不用"simultaneously"或"at the same time"。（3）名词"effect"的搭配模式是"the effect of A on[upon] B"。（4）为了使英语句子紧凑，所以第二句应合并到第一句中去，采用非限制性定语从句的形式，即写成"on the basis of which…"。（5）"kind of"应该去掉。（6）表示"……的表示式"一般应使用"an expression for…"为好。

**【改正后的句子】**　Also, a quantitative expression for the effect of all these factors on the performance of the system is derived, on the basis of which an improved system is presented.

### 例 86

**【汉语原句】**　该设备能够检测集成电路的好坏以及鉴别无名芯片的类型。

**【英语错句】**　This device can detect whether an IC is good or bad and identify the type of a nameless chip.

**【错误分析】**　（1）这儿的"好坏"不能简单地表示为"good or bad"，它的真实意思是"能否正常工作"，所以应该表示成"function properly or not"，即根据技术指标来确定的。（2）这儿的"无名"，并不是真正的没有名字，而是我们不知道而已，所以不能用形容词"nameless"，而应该用形容词"unknown"才是确切的。

**【改正后的句子】**　This device can detect whether an IC functions properly or not［…detect the quality of an IC］and identify the type of an unknown chip.

**例 87**

【汉语原句】 当且只有当以下两个命题中的任何一个成立时,强完美图猜想成立。

【英语错句】 The Strong Perfect Graph Conjecture is true if and only if any of the following two propositions is true.

【错误分析】 (1)这儿的"强"是修饰"完美"的,因此应该用副词"strongly",这种错误经常发生,又如"这是一个任意给定的数"应写成"This is an arbitrarily given number."而不少读者往往会使用形容词"arbitrary"。(2)由于在此涉及两个命题,所以这儿的"任何一个"只能用"either",或者使用"one or the other"。

【改正后的句子】 **The Strongly Perfect Graph Conjecture is true if and only if either of the following two propositions is true.**

**例 88**

【汉语原句】 这种新的算法保持了能很快找到最佳路线的优点。

【英语错句】 This kind of new algorithm remains the advantage of finding the optimal routes very quickly.

【错误分析】 (1)"kind of"应该去掉。(2)"remain"只能当连系动词或不及物动词,它并不是及物动词,在此只能用"maintain"来代替它。要注意,当"remain"译成"保持"时,它是连系动词,后面只能跟形容词或名词作它的表语。

【改正后的句子】 **This new algorithm maintains the advantage of finding the optimal routes very quickly.**

**例 89**

【汉语原句】 这一过程非常类似于流质从一根脉管经过神经网络流到另一根脉管的情况。

【英语错句】 This process is much similar to that a fluid flows through a neural network from a vessel to another.

【错误分析】 (1)"much"一般不能修饰形容词,应把它改成"very"。(2)"that"不能引导介词宾语从句,在此根据科技概念应改成"the one by[in] which","one"在此表示"process"。(3)当跟"another"连用时不能用不定冠词"a"或"an"来表示"一"的含义,应该使用数词"one"作定语。

【改正后的句子】 **This process is very similar to the one by which a fluid**

**flows from one vessel through a neural network to another.**

**例 90**

【汉语原句】 在有效区内场的变化小于 4 dB，可以满足电磁兼容测量的要求。

【英语错句】 The variation of the field in effective region is smaller than 4dB. This shows that the demands in electromagnetic compatibility can be satisfied.

【错误分析】 (1) 在"effective"前应该加上定冠词。(2) 表示数值的"小"的比较级，应使用"less"而不用"smaller"。(3) 英语文本的两个句子应该用一个句子来表示，这样就显得紧凑，即"…, which can satisfy the requirements for electromagnetic compatibility measurement"，注意技术上的"要求"应该使用"requirement"一词而不是"demand"。

【改正后的句子】 **The variation of the field in the effective region is less than 4 dB, which can satisfy the requirements for electromagnetic compatibility measurement.**

**例 91**

【汉语原句】 解决了硅片和封装玻璃之间热膨胀系数不一致的难题。

【英语错句】 The difficulty of the difference of the thermal expansion coefficients between silicon substrate and packaging glass is settled.

【错误分析】 (1) "难题"一般表示成"the difficult problem"。(2) "不一致"为"inconsistency"。(3) "A 和 B 在 C 方面的……"，应采用"…in C between A and B"这一表示法，希望读者能够熟悉它。(4) "解决"一般用"solve"来表示（也有人用"resolve"的）。

【改正后的句子】 **The difficult problem of inconsistency in thermal expansion coefficient between silicon substrate and packaging glass is solved.**

**例 92**

【汉语原句】 在讨论了 LoG 滤波器及其设计后，详细地描述了如何在 FPGA 中通过 LoG 滤波来实现红外目标检测。

【英语错句】 After the discussion of LoG filter and its design, detail description on how to realize the infrared image target detection in FPGA through LoG filtering is followed.

【错误分析】 (1) 在"LoG filter"前应该有定冠词"the"，因为它在这里是可数名词单数。(2) "detail description"应该改为"a detailed description"，注意不定冠词的特殊用法。(3) 在"description"后应该用"of"引出它的逻辑宾语。(4) "is followed"意为"被跟着"，这里应该使用一个作为不及物动词的"follow"，意为"随之(跟随)而来"。

【改正后的句子】 **After the discussion of the LoG filter and its design，a detailed description of how to realize the infrared image target detection in FPGA through LoG filtering follows.**

例 93

【汉语原句】 这算法与以往的方法相比有两个优点：输入数据量少，可以任意更改地形的局部特征。

【英语错句】 Compared with the existing algorithms, this algorithm has two advantages that it needs less input data and local characteristic of terrain can be changed artificially.

【错误分析】 (1) 凡遇到"advantage"时，表示"与……相比"之意一般应该使用"over"。(2) "less"应该改成"fewer"，因为它修饰的是一个可数名词复数"data"。(3) "artificially"的意思是"人工地"，而"任意"应该是"arbitrarily"表示。(4) 凡有几个并列的从句就应该使用几个从句引导词，所以在"data"后应该加上同位语从句引导词"that"。(5) 还有两处没有加定冠词。

【改正后的句子】 **This algorithm has two advantages over the existing ones that it needs fewer input data and that the local characteristic of the terrain can be changed arbitrarily.**

例 94

【汉语原句】 我们假设该系统中部件工作时间服从指数分布，修理时间是连续分布，并且系统中的部件存在两种失效模式。

【英语错句】 It is assumed that the working time of the components in the system is exponentially distributed, response time of these components is continually distributed, and the components in the system are subject to two different kinds of failure.

【错误分析】 (1) 由于假设的内容有三个，根据语法规则应该用三个"that"分别引出三个主语从句。(2) 在"response time"前应该加有定冠词，因为它带有"of"引出的后置定语。(3) 本句中涉及的"连续"的含义是"不间断地”，所以应该使用“continuously”而不是“continually”（这个词的词意为 constantly and frequently recurring；always happening）。（4）句中的“kinds”应该改成“modes”。

【改正后的句子】 **It is assumed that the working time of the components in the system is exponentially distributed，that the response time of these components is continuously distributed，and that the components in the system are subject to two different modes of failure.**

例 95

【汉语原句】 最后获得了 HIV/AIDS 的平均费用相当于总财政的情况。

【英语错句】 Eventually，the average cost for HIV/AIDS vs. total finance is obtained.

【错误分析】 （1）“Eventually”与“at last”和“in the end”的含义雷同，意为“最终”，所以这里应该使用“finally”一词。（2）根据本句及文章里的具体情况，句子的主语应该是“a curve of …”来表示“情况”。（3）在“average cost”前应该加定冠词。（4）这里的“总财政”应该表示为“the total amount of finance”。

【改正后的句子】 **Finally，a curve of the average cost for the HIV/AIDS vs. the total amount of finance is obtained.**

例 96

【汉语原句】 本文提出了一种改进方法，它可以加速信噪分离。

【英语错句】 This paper makes an improving method. It can accelerate the separation of the signal and the noise.

【错误分析】 （1）表示“提出”应该使用“presents”或“proposes”。（2）这里的“改进”应该使用过去分词来表示。（3）为使得句子紧凑，把两句中的句号改成逗号，把“It”改成“which”。（4）表示“A 与 B 分离”应该使用 “the separation of A from B”这一搭配模式。

【改正后的句子】 **This paper presents an improved method，which can accelerate the separation of signals from noise.**

例 97

【汉语原句】 对解析表达式与模拟结果作了比较，表明它们是很吻合的。

【英语错句】 The comparison of the analytical expression with simulation results is given and both are matched well.

【错误分析】 (1)"作比较"应该使用"make a comparison"。(2)本句第二部分应该用一个非限制性定语从句把前后两者紧密联系起来,即"which shows that …"。(3)这里的"吻合"并不是"匹配"的意思,所以不能使用动词"match"一词,而应该使用"agreement"来表示,即"they are in good agreement"。

【改正后的句子】 **A comparison of the analytical expression with simulation results is made,which shows that they are in good agreement.**

**例 98**

【汉语原句】 LED 装置包括许多高功率的 LED,其亮度是可以调整的。

【英语错句】 LED unit includes many high-power LED and its lightening can be adjusted.

【错误分析】 (1)句子开头的"LED"之前应该有定冠词,因为这里显然是特指。(2)句中的"包括"的确切含义应该用"contain"来表示。(3)句中第二个"LED"后应该用复数形式。(4)"亮度"应该是"brightness""luminance;brilliance"等,而"lightening"的意思是"闪电放电"。(5)第二个分句用"with 结构"来表示就能使整个句子显得紧凑,即"with its brightness being adjustable"。

【改正后的句子】 **The LED unit contains many high-power LEDs,with its brightness(being)adjustable.**

**例 99**

【汉语原句】 而环境质量在不同的区域是很不相同的。

【英语错句】 While the environment quality differs greatly by different districts.

【错误分析】 (1)这里的"而"绝不能用"while"来表示,因为它是一个状语从句引导词,句中一定要有主句和从句两部分,这也是不少读者常犯的一个错误,应该使用"however";如果有主句和从句,则"while"一般处于主句后表示对比,这时它可表示"而"之意,如:"Transistor Q1 is on, while transistor Q2 is off.(晶体管 Q1 导通,而晶体管 Q2 截止。)"。(2)"环境质量"应该是"the environmental quality"或"the quality of environment",这与"数学计算(mathematical calculation)""实验结果(experimental results)""理论分析(theoretical analysis)"等表达形式是类同的,要用形容词来作定语。(3)"不同的区域"在此应该写成"in different regions"。

【改正后的句子】 **However,the environmental quality differs greatly in** different regions.

**例 100**

**【汉语原句】** S 模块的作用由如下三个方面构成。

**【英语错句】** The job of S-module is consisted of the three aspects as follows.

**【错误分析】** （1）这里的"作用"应该用"function"来表示。（2）在"S-module"前应该有定冠词，因为它是一个特定的东西。（3）"consist of"属于主动的形式、被动的含义，没有被动形式，这也是一些读者常犯的错误。（4）"as follows"这个固定的短语只能作状语或表语，它是不作后置定语的。例如：

This phenomenon can be explained **as follows**.

这个现象可以解释如下。

In this case, we can express Ohm's law **as follows**.

在这种情况下，我们可以把欧姆定律表示如下。

The block diagram of a digital computer is **as follows**.

数字计算机的方框图如下。

所以要作后置定语的话，应该写成"the three aspects which follow"或"the following three aspects"。

**【改正后的句子】** **The function of the S-module consists of the following three aspects.**

**例 101**

**【汉语原句】** 这部分是用来回答市民询问不同地点路况信息的。

**【英语错句】** This part is used to reply the messages sent by citizens enquiring about the traffic informations of different locations.

**【错误分析】** （1）"reply"应该改成"answer"。（2）"the messages sent by citizens enquiring"应该改为"the enquiries from citizens"。（3）"information"是个不可数名词，后跟"on"或"about"。（4）"路况"'应该是"road conditions"。（5）"of"改成"in"。

**【改正后的句子】** **This part is used to answer the enquires from citizens about the information on road conditions in different locations.**

**例 102**

**【汉语原句】** 该设备适用于装有 SmartLight 系统的住户。

【英语错句】　The device is suit to households that equip SmartLight system.

【错误分析】　(1)"suit"是个动词，有些读者经常把它用错，在此应该使用形容词短语"suitable for"。(2)"装有……"的表示法应该为"be equipped with"。(3)在"SmartLight system"前应该加上定冠词"the"。

【改正后的句子】　**The device is suitable for households（that are）equipped with the SmartLight system.**

**例 103**

【汉语原句】　图 1 画出了可能会产生具有低 PAPR 的序列的情况。

【英语错句】　Fig. 1 draws the case of generating a sequence with low PAPR possibly.

【错误分析】　(1)"draws"应该改为"shows"。(2)"the case … possibly"改为"the possibility"。(3)在"low"之前应该加上冠词"a"。

【改正后的句子】　**Fig. 1 shows the possibility of generating a sequence with a low PAPR.**

**例 104**

【汉语原句】　实验结果表明，MBNM 是一种很有希望的方法，因为它具有高的分类精度和低的维数。

【英语错句】　The experimental results show that MBNM is a kind of promising method due to high classification accuracies and low dimensions.

【错误分析】　(1)在文中没有提到做实验的情况下，去掉句首的定冠词。(2)在"MBNM"前要加上"the"。(3)去掉"kind of"。(4)在"due to"后加上"its"。

【改正后的句子】　**Experimental results show that the MBNM is a promising method due to its（leading to）high classification accuracies and low dimensions.**

**例 105**

【汉语原句】　一个良好的距离尺度可以为我们对构成数据结构提供某些启示。

【英语错句】　A good distance metric may provide us some enlightenment in underlying data structures.

【错误分析】　(1)"provide"的搭配关系是"provide sb with sth"。

（2）"enlightenment"后跟"on"。（3）"underlying"应改为"constructing"。

**【改正后的句子】** A good distance metric may provide us with some enlightenment on constructing data structures.

### 例 106

**【汉语原句】** 然后，对实验结果作了有关的分析。

**【英语错句】** Then related analysis on experiment results is followed.

**【错误分析】** （1）在"Then"后加上冠词"a"，在"分析（analysis）""研究（study）""检查（examination；inspection）""比较（comparison）""考虑（consideration）"等名词前往往加不定冠词。（2）"on"改成"of"引出"analysis"的逻辑宾语。（3）"experiment"应改成"experimental"。（4）"is followed"意为"被跟着"，这里应该改成不及物动词"follows"，表示"随之而来了"。

**【改正后的句子】** Then a related analysis of experimental results follows.

### 例 107

**【汉语原句】** 由于 SVDD 核形式过于简单，把它由单个高斯核扩展为多个高斯核线性组合形式。所获得的多核支持向量域描述可以表示为半正定规则问题。

**【英语错句】** Owing to the disadvantage of having too simple kernel formation，SVDD is expanded from a single Gaussian kernel to a linear combination of multi-Gaussian kernels. The resulted Multi-kernal SVDD could be expressed as a SDP problem.

**【错误分析】** （1）把"Owing to"改为"Due to"，因为前者一般用于好事。（2）由于"formation"在此是可数名词单数，表示"方式"，所以在"kernel"前加"a"，而由于"too"的关系，要采用不定冠词的特殊位置，即"too small a…"。（3）在"SVDD"前加"the"。（4）"resulted"应改成"resulting"或"resultant"，因为动词"result"是不及物动词，这是一些读者常犯的错误。（5）"SDP"由于其读音的关系而要把其前面的"a"改为"an"。

**【改正后的句子】** Due to the disadvantage of having too simple a kernel formation，the SVDD is expanded from a single Gaussian kernel to a linear combination of multi-Gaussian kernels.The resulting Multi-kernel SVDD could be expressed as an SDP problem.

### 例 108

**【汉语原句】** 这些算法难以保持原图像的边缘和细节。

【英语错句】　These algorithms are difficult to preserve edges and details of the original image.

【错误分析】　这里不能使用"are difficult to"，这是受到汉语的影响而出现的错误，不能有"人（物）is［are］difficult［easy］to do sth."而应该在句尾加上"with difficulty［ease］"。但最好写成"It is difficult［easy］for 人［物］to do sth."。

【改正后的句子】　**These algorithms preserve edges and details of the original image with difficulty.**（或：**It is difficult for these algorithms to preserve edges and details of the original image.**）

例 109

【汉语原句】　约束参数选取得越大，性能改善越接近于最优。

【英语错句】　The constraint parameter is selected as more larger, the performance improvement will be more closer to the optimization.

【错误分析】　（1）"越……，越……"句型应该是"The＋比较级…，the＋比较级…"，所以本句句首应该写成"The larger the constraint parameter is selected to be，"。（2）逗号后应该以"the closer to optimization"开头。（3）去掉"more"。（4）去掉"optimization"前的"the"。

【改正后的句子】　**The larger the constraint parameter is selected to be，the closer to optimization the performance improvement will be.**

例 110

【汉语原句】　本文提出了一种视频鲁棒性传输系统，在此基础上提出了 SVC 特性描述方式。

【英语错句】　This paper puts forward a video robustness transmission system，based on which the SVC characteristics description framework is proposed.

【错误分析】　（1）本句主要错在"based on which"，应改为"on the basis of which"，这也是有些读者常犯的错误。（2）"characteristics"应该用单数形式。

【改正后的句子】　**This paper puts forward a video robustness transmission system，on the basis of which the SVC characteristic description framework is proposed.**

例 111

【汉语原句】　应当注意，这两个信道的频率也不可能均衡得超过$-f_s/2\sim$

$f_s/2$。

【英语错句】 It should be given regard that the frequencies of the two channels can't be equalized beyond$-f_s/2\sim f_s/2$ also.

【错误分析】 (1)"given regard"应该改成"noted"。(2)在科技文中,一般不用单词的紧缩形式,所以"can't"应该写成"cannot"。(3)由于主语从句是一个否定句,所以"also"应该改成"either"。

【改正后的句子】 **It should be noted that the frequencies of the two channels cannot be equalized beyond$-f_s/2\sim f_s/2$ either.**


**例 112**

【汉语原句】 把同样的测试信号,例如脉冲和线性 FM 信号,加到每个通道上。

【英语错句】 The same test signals,for example pulse and linear FM signals,are added to every channel.

【错误分析】 (1)在表示一个一个事物时的"例如"通常使用"such as"而不是"for example"。(2)加信号的"加"应该使用"apply"而不是"add",有些读者会经常弄错。

【改正后的句子】 **The same test signals,such as pulse and linear FM signals,are applied to every channel.**


**例 113**

【汉语原句】 $Cl_2/Ar$ 中加入 $O_2$ 对 GaN 刻蚀速率影响不大,却使$Al_{0.27}Ga_{0.73}N$ 刻蚀速率明显下降。

【英语错句】 Adding $O_2$ to $Cl_2/Ar$ effects little etch rates of GaN,but etch rates of $Al_{0.27}Ga_{0.73}N$ reduced obviously.

【错误分析】 (1)表示"影响"的动词是"affect"而不是"effect",这是不少读者经常出错的。(2)"but"之前的谓语和宾语应该表示成"has a little effect on the etching rate of GaN"。(3)"but"之后可以写成"leads to a significant reduction in the etching rate of …"。

【改正后的句子】 **Adding $O_2$ to $Cl_2/Ar$ has a little effect on the etching rate of GaN,but leads to a significant reduction in the etching rate of $Al_{0.27}Ga_{0.73}N$.**


**例 114**

【汉语原句】 挖掘算法将节点划分到不同的集合中,每个集合展现系统的一个设计侧面。

【英语错句】　The mining algorithm separates the vertices into different subsets and each subset presents one design aspect of the system.

【错误分析】　（1）"separate … into …"意思是"把……分解成……"，这里应该使用"allot/allocate … to …"意为"把……分配到……"。（2）"and"及其后面的东西应该用一个"with 结构"来表示，使句子显得紧凑。

【改正后的句子】　**The mining algorithm allots/allocates the vertices to different subsets，with each subset presenting one design aspect of the system.**

例 115

【汉语原句】　根据最小逼近误差原则，我们寻找几何方向。

【英语错句】　Based on the minimal approximation error，we search the geometry directions.

【错误分析】　（1）在"on the"后面最好加上"principle of"。（2）"search"意为"搜查"，而"寻找"应该用"search for"。（3）应该把"geometry"改成"geometrical"，同时把其前面的定冠词去掉，因为该词是复数形式且没有特指。

【改正后的句子】　**Based on the principle of minimal approximation error，we search for geometrical directions.**

# Ⅲ、词的搭配

例 116

【汉语原句】　获得的算法被称为 MOEA/DD。

【英语错句】　The resulted algorithm is termed as the MOEA/DD.

【错误分析】　（1）"resulted"应该改为"resulting"，因为"result"作动词时为不及物动词。（2）要去掉介词"as"，使用"term"或"call"或"name"表示"把……称为……"时不得使用介词"as"，不少读者常犯这一错误。

【改正后的句子】　**The resulting algorithm is termed the MOEA/DD.**

例 117

【汉语原句】　地球的这些信息是非常有用的。

【英语错句】　These informations of the earth are very useful.

【错误分析】 （1）"information"是一个不可数名词，不能有复数形式，可以用"these pieces of information"来表示复数。（2）"information"要搭配介词"on"或"about"。

【改正后的句子】 **These pieces of information on the earth are very useful.**

### 例 118

【汉语原句】 分析了轨道状态的特性，构建了轨道变化中共面变轨和异面变轨的等级划分方法。

【英语错句】 The orbit status characteristics is analyzed，and the hierarchical division method of coplanar and noncoplanar orbit change are constructed.

【错误分析】 （1）由于"characteristics"是一个复数名词，所以其后面要把"is"改成"are"。（2）在此"method"后要用介词"for"。（3）在"coplanar"前要加冠词"the"。（4）"are"要改成"is"。

【改正后的句子】 **The orbit status characteristics are analyzed，and the hierarchical division method for the coplanar and noncoplanar orbit change is constructed.**

### 例 119

【汉语原句】 对于原来的分解方法的另一个改进是减少改进区域的面积。

【英语错句】 Another improvement to the original decomposition method is to reduce the volume of the improvement region.

【错误分析】 （1）在"improvement"后跟"over""in""on""upon"而不跟"to"；（2）"面积"应该是"area"，而"volume"是"体积"。（3）后面的"improvement"应该改成"improved"。

【改正后的句子】 **Another improvement over the original decomposition method is to reduce the area of the improved region.**

### 例 120

【汉语原句】 这是圆的方程。

【英语错句】 This is an equation of a circle.

【错误分析】 （1）把"an"改成"the"。（2）"equation"后面应该跟"to"或"for"。同样的有"the answer to/for …""the solution to/for …""the symbol for …"。有的读者往往加"of"，搭配不对。

【改正后的句子】 **This is the equation to/for a circle.**

**例 121**

【汉语原句】 给出了 $k$-错线性复杂度严格小于线性复杂度的一个充分必要条件。

【英语错句】 A necessary and sufficient condition for which the $k$-error linear complexity is strictly less than the linear complexity is showed.

【错误分析】 （1）在"condition"之后一般是跟"for"短语或一个同位语从句，而不用定语从句。（2）根据观察，在"condition""constraint""restraint"等名词后的同位语从句和表语从句中，多数英美科技人员使用"(should)＋动词原形"型虚拟语气句型。例如：

A necessary **condition** that this be the case is that Euler's Equation be satisfied.

出现这一情况的一个必要条件是：要满足尤拉方程。

This might have been expected because of the **constraint** that there be no additional power supplied by the source.

这一点可能早该预料到了，因为有这样的制约条件：电源不再提供能量了。

（3）动词"show"的过去分词多数用的是"shown"，极少用"showed"。

【改正后的句子】 **A necessary and sufficient condition that the $k$-error linear complexity（should）be strictly less than the linear complexity is shown.**

**例 122**

【汉语原句】 该电阻上的电压为零点几伏特。

【英语错句】 The voltage on the resistor is zero point several volt.

【错误分析】 这个句子属于典型的中文式英语句：（1）表示电压"voltage"、电容"capacitance"等其后一般使用介词"across"而不用"on"，因为它们都是在两点之间形成及测量的。（2）关于"零点几""零点零几"［即"十分之几""百分之几"］等表达法应该采用分数表示方法，其公式为：

**"a few/several［这是分子］＋tenths（hundredths 等等）［这是分母］＋of＋a/an＋单位"**

如："零点几伏特"应表示为"a few（或 several）tenths of a volt"；"零点零几安培"应表示成"a few（或 several）hundredths of an ampere"。

【改正后的句子】 **The voltage across the resistor is a few tenths of a volt.**

**例 123**

【汉语原句】　这时候晶体管 Q1 导通，而晶体管 Q2 截止。

【英语错句】　On this point transistor Q1 is on；while transistor Q2 is off.

【错误分析】　（1）表示某一时刻时，应该用介词"at"。（2）"while"表示"而"的含意时仍属于"从属连词"而不是"并列连词"，所以其前面不能用分号（分号表示一个并列连词的含意），但可以有逗号。当然这里也可以把"while"改用副词"however"。有的读者甚至会写成独立的一句"While transistor Q2 is off."这时要把"While"改成"However"。对于"while"表示含意"而"时的句型在此举几个例子（这时主从句中的句型一般是相同的）：

Water is a liquid，**while** ice is a solid.

水是液体，而冰则是固体。

Input A goes low **while** input B stays high.

输入 A 变成了低电位而输入 B 则保持高电位。

At this time the kinetic energy approaches infinity，**while** the potential energy approaches the minimum.

这时动能趋于无穷大，而势能则趋于最小值。

**While** energy is the capacity to do work，power is the quantity of work in unit time.

能量是做功的能力，而功率是单位时间内所做功的数量。——为了强调对照，从句放在主句之前了，可译成汉语时仍应把"而"放在第二个分句前：

**While** the heat of a body indicates the combined energies of all of its molecules，the temperature of a body measures the average of each individual molecule.

物体的热量表示了其所有分子的能量之和，而物体的温度则度量每个分子的平均能量。

【改正后的句子】　**At this point[time；moment] transistor Q1 is on，while transistor Q2 is off.**

**例 124**

【汉语原句】　控制 Mask 微形变化的方法研究

【英语错句】　A study on the control method of mask of very small distortion

【错误分析】　这是一篇论文的标题。（1）我国杂志机构要求文章标题的第一个冠词应该省去。关于论文标题中的字母大写问题，目前国内外学术期刊上有两种写法：① 整个标题的第一个字母要大写，其余的字母均小写，这是目前的趋势；② 整个标题的第一个字母及后面每个实词的第一个字母均要大写，这是过去沿袭下来的写法。(2) 名词"study"来自于及物动词，所以其后多数人使用"of"来引出它的逻辑宾语（也有人用"on"的）。(3) "control"的对象是"Mask 微型变化"，它们应该紧靠在一起。(4) 在"method"后常见跟"for"，在此写成"for controlling the very small distortion of the Mask"。

【改正后的句子】 **Study of the method for controlling the very small distortion of the mask（或：···for the control of the very small distortion of the mask）**

例 125

【汉语原句】 这是由于在 PN 结上存在一个电容之故。

【英语错句】 This is due to that there exists a capacitance on the PN junction.

【错误分析】 (1) "that"是不能引导介词宾语从句的（除了"in that"表示"因为""在于"和"except that"表示"除……之外"这两个固定词组外），为了避免这个问题，英美人一般在"that"之前加上"the fact"两词。这是不少读者，甚至有的英语教师，并不了解这一点。(2) 介词"on"在此用错了，因为电容是在两块平板间形成的，所以要用介词"across"一词。

【改正后的句子】 **This is due to the fact that there exists a capacitance across the PN junction.（或：This is due to the existence of a capacitance across the PN junction.）**

例 126

【汉语原句】 这台转动的机器将在一两分钟后自动停下来。

【英语错句】 This running machine will stop of itself after one minute or two.

【错误分析】 (1) 在表示瞬间动作或状态的动作将来时的句中要表示"在……之后"之意时要用"in"而不能用"after …"或"… later"。(2) 表示"一两天""一两分钟""一两个月"等等概念时有两种表达方式，一种是"one or two days [minutes; months]"，另一种是"a day [minute; month] or two"。

【改正后的句子】 **This running machine will stop of itself in one or two minutes [ in a minute or two].**

**例 127**

【汉语原句】　在 10 月 8 日早上又发射了一颗通信卫星。

【英语错句】　Another comsat has been launched in the morning of the 8th of October.

【错误分析】　(1) 时态用错了，由于本句具有明显的过去时状语"10 月 8 日早上"，所以只能用一般过去时，这种情况经常被一些读者误解而用了现在完成时，主要受到了汉语"了"字的影响。(2) 当"上午""下午"等搭配有特定日期作后置定语时，根据语法规则就不能再用介词"in"了，而必须用"on"。

【改正后的句子】　**Another comsat was launched on the morning of the 8th of October.**

**例 128**

【汉语原句】　该实验是从八点钟开始的。

【英语错句】　The experiment began from eight o'clock.

【错误分析】　介词"from"在此用错了，主要受到了汉语的影响，应该使用"at"。同样若表示"从七月份开始"应该用"in July"；"从明天开始"应使用"tomorrow"一词。

【改正后的句子】　**The experiment began at eight o'clock.**

**例 129**

【汉语原句】　电压是用伏特来度量的。

【英语错句】　Voltage is measured with volt.

【错误分析】　(1) 这里的"用"不该用"with"，英美人均用"in"来表示。(2) "volt"这样的单位均要用复数形式(只有"hertz"除外，只用单数，主要是复数形式不好构成)。

【改正后的句子】　**Voltage is measured in volts.**

**例 130**

【汉语原句】　通过传统的循环调试策略，设立断点或单步调试就可定位错误。

【英语错句】　Through the traditional circulation debugging strategy, setting the breakpoint or debugging step by step to locate the error.

【错误分析】　(1) 这里的"通过"实际上意为"根据"，所以要把"Through"改为"By"。(2) 逗号后不是完整句，因此要把"setting"和"debugging"改为"it is possible to set …and debug …"。

**【改正后的句子】** By the traditional circulation debugging strategy，it is possible to set the breakpoint or to debug step by step to locate the error.

### 例 131

**【汉语原句】** 通过使用统计方法，本文提出了在优化过程中不必考虑的单元数目的计算公式。

**【英语错句】** Through using statistical method，this paper presents the calculation method of the number of elements which doesn't need to be considered in the optimization process.

**【错误分析】** （1）把"Through"改为"By"，这时"by"就意为"通过"，在"through"后一般不跟动名词，可以跟"名词＋of …"，所以这里也可用"Through the use of the statistical method"。（2）在"using"后加"the"。（3）"the calculation method of"应改为"the method for calculating"。（4）在科技文中"doesn't"应该写成"does not"。

**【改正后的句子】** By using the statistical method，this paper presents the method for calculating the number of elements which does not need to be considered in the optimization process.

### 例 132

**【汉语原句】** 通过分析电容耦合连接器物理结构对接收端脉冲波形的影响，提出了连接器尺寸的选择策略。

**【英语错句】** Through analyzing the effect of the physical structure of the capacitive coupling connector on pulse waveform in the receiving end，a selection strategy of the size of the connector is proposed.

**【错误分析】** （1）与上句一样，要把"Through"改成"By"。（2）在"pulse"前加"the"。（3）把"waveform"后的"in"要改成"at"。（4）把"a selection strategy of"改成"a strategy for selection of"或"for selecting"。

**【改正后的句子】** By analyzing the effect of the physical structure of the capacitive coupling connector on the pulse waveform at the receiving end，a strategy for selecting the size of the connector is proposed.

### 例 133

**【汉语原句】** 同时，该方法甚至在低 SNR 时能获得精确的识别。

**【英语错句】**　As well as the method can obtain accurate recognition even at low SNR.

**【错误分析】**　（1）这里的"同时"并不表示时间，实际上它表示"另外"之意，一般用"Also"来表示。"As well as"放在句首意为"除……外（还……）"。（2）应该把"obtain"改成"lead to"，因为"方法"自己不会"得到"的，这也是有些读者常犯的错误。（3）把"at"改成"for a"。

**【改正后的句子】**　**Also，the method can lead to accurate recognition even for a low SNR.**


例 134

**【汉语原句】**　这个反馈信号可多次被使用。

**【英语错句】**　This feedback signal can be used for many times.

**【错误分析】**　可能是原作者把"many times"看成是时间了，所以在其前面使用了介词"for"，实际上它在此表示的是"次"，所以只能使用介词"by"。在英语中除了表示"时间"和"距离"要使用介词"for"（表示距离还可使用"through"）外，表示其它数量时一般都要用介词"by"（表示"温度"和"角度"时也可使用介词"through"），这个介词也是可以省去的。

**【改正后的句子】**　**This feedback signal can be used（by）many times.**


例 135

**【汉语原句】**　由于无线传感器网络的特殊性，针对其节能问题的媒体访问控制（MAC）协议研究越来越受到人们的关注。

**【英语错句】**　Because of the particularity of wireless sensor network，the research of MAC protocol is gaining more and more attention as a main approach to maximize the network efficiency.

**【错误分析】**　（1）在"wireless"前加"the"。（2）在"MAC"前加"the"。（3）要把"research"后的"of"改成"on"，在"research"后常见跟介词"on"，它也可以跟"for""with""into""in"，但不能跟"of"。（4）把"gaining"改成"attracting"，当主语是"事物"时，与"attention"连用的动词有"attract""draw""catch"等。（5）把"to maximize"改成"to maximizing"，因为这里"approach"后的"to"是介词。

**【改正后的句子】**　**Because of the particularity of the wireless sensor network，the research on the MAC protocol is attracting more and more attention as a main approach to maximizing the network efficiency.**

例 136

【汉语原句】　定义了一种描述图论中某些工程问题求解的算法语言(我们称之为图论算法语言)。

【英语错句】　An algorithm language (we call it the Algorithm Language for the Graph Theory) to describe the solution of some engineering problems in the graph theory is defined.

【错误分析】　(1)应使括号内的东西与括号外紧密相关,所以在"we"之前应加上关系代词"which"而去掉括号内的代词"it"。(2)"to describe"应改成"for describing",在"Language"后跟介词"for",表示前面的算法语言是用于什么目的的。

【改正后的句子】　**An algorithm language (which we call the Algorithm Language for the Graph Theory) for describing the solution of some engineering problems in the graph theory is defined.**

例 137

【汉语原句】　本文介绍了应用正向-反向线性预测来提高距离校准的精度的方法。

【英语错句】　A method applying the forward-backward linear prediction to increase the accuracy of range alignment is presented.

【错误分析】　(1)本句的错误在于"applying…"是修饰"to increase the accuracy…"的方式状语,所以应该通过"by 短语"放在不定式短语的后面才对。(2)不少英美人在"method"后喜欢用介词"for"。

【改正后的句子】　**A method for increasing the accuracy of range alignment by applying the forward-backward linear prediction is presented.**

例 138

【汉语原句】　本文引入所谓的网络"站级"概念,并基于此概念将网络中的所有结点在源-宿方向上划分成等级,从而构成不同等级的结点集。

【英语错句】　A so-called "node-step" concept is introduced. Based on this concept, all nodes in the network are organized into steps on the source-to-destination direction so as to form several sets of stepped nodes.

【错误分析】　(1)"所谓……概念"正规的表示方法应该是"the concept of so-called…"。(2)在方向"direction"之前只能用介词"in",而绝不能受汉语的影响用"on"或"to"或"toward"。(3)在汉语句中第一个逗号前后的两部分内容中有一个词"概念"是共有的,因而应该使用定语从句来表示第一个逗号后的那部分内容而使英语句子显得紧凑,在构成定语从句时,不能写成"based on which",而应该写成"on the basis of which"。

**【改正后的句子】** The concept of so-called "node-step" is introduced, on the basis of which all nodes in the network are organized into steps in the source-to-destination direction so as to form several sets of stepped nodes.

**例 139**

**【汉语原句】** 某些元素产生放射的能力被称为放射性。

**【英语错句】** The ability for some elements to give off radiations is referred as the radioactivity.

**【错误分析】** (1) 在前面的例句中曾讲到,在"ability"和"tendency"等抽象名词后一般应该采用"of"来引出不定式复合结构作其后置定语(见到只有极少数人用"for"的),所以在此应该把"for"改为"of",下面再举几个例子:

A few factors affect the ability **of a capacitor to store charge.**

有好几个因素影响电容器贮存电荷的能力。

The ability **of a receiver to discriminate between signals of different frequencies** is called receiver selectivity.

接收机区别不同频率的信号的能力就称为接收机的选择性。

The greater the tendency **of an object to resist a change of velocity**, the greater its inertia.

物体阻止速度变化的趋势越大,其惯性就越大。

Elasticity may be defined as the tendency **of a body to return to its original state after being deformed.**

弹性可以被定义为物体在形变后恢复其原状的趋势。

(2) 表示"被称为"有好几种表示方法,如"be called …""be termed …""be named …""be known as …""be referred to as …""be spoken of as …"等,本句中在"referred"后面漏了介词"to",在主动式中,应该为"refer to A as B"(把 A 称作为 B),变成被动句后介词"to"必须留在过去分词"referred"之后。

(3) "radioactivity"(放射性)在此为泛指的抽象名词,所以其前面一般不用定冠词"the"。

**【改正后的句子】** The ability of some elements to give off radiations is referred to as radioactivity.

**例 140**

**【汉语原句】**　这种方法在一定程度上能够解过拟合问题。

**【英语错句】**　This kind of method can solve the overfitting problem in certain extent.

**【错误分析】**　(1) 应该去掉"kind of"。(2) 在表示"程度"的名词"extent""degree""point"前常用介词"to"，而如果使用"measure"来表示"程度"的话，在其前面就要使用介词"in"了，这是词汇搭配问题。(3) 在"certain extent"前应该有不定冠词"a"。

**【改正后的句子】**　**This method can（be used to）solve the overfitting problem（或：overfitting problems）to a certain extent.**

**例 141**

**【汉语原句】**　直流电总是朝一个方向流动。

**【英语错句】**　The direct current always flow to one direction.

**【错误分析】**　(1) 由于主语为单数第三人称，所以"flow"后应该加"s"，这是初始学习科技英语写作的人经常犯的一个错误。(2) 表示"朝……方向"时应该用"in"，有些英语名词前要用特定的介词，如"在这个温度上"要表示成"at this temperature"，"为此目的"可表示成"for this purpose"或"to this end"（虽然词典上说在"end"前也可用"for"，但英美人在实际应用时主要用"to"），有些英语名词后要用特定的介词，如"对……的影响"要表示成"the effect [influence；impact] on [upon]"，"与……的关系"要表示成"the relationship to …"，"物理实验"一般表示成"an experiment in physics"。(3) "直流电"在此仅表示一个概念，在其前面一般不使用定冠词。(4) 由于受汉语影响中国人往往把"总是(always)"这个副词放在动词之前，而英美人在这种情况下往往把它放在介词短语之前，侧重点在后面，又如："这台机器主要由五部分组成。"应译成"This machine consists **mainly** of five parts."，不应把"mainly"放在"consists"之前，再如："这个定律只适用于金属导体。"应译成"This law applies **only** to metallic conductors."。

**【改正后的句子】**　**Direct current flows always in one direction.**

**例 142**

**【汉语原句】**　除了该电子外，氢原子还含有一个正质子。

**【英语错句】**　Except the electron, hydrogen atom also contains a positive proton.

【错误分析】　（1）本句中表示的"除……以外"从全句含义看是其本身也包括在内的，所以只能用"besides"或"in addition to"来表示。（2）"氢原子"是可数名词，其前面必须加上冠词。

【改正后的句子】　**In addition to the electron, a hydrogen atom also contains a positive proton.**

**例 143**

【汉语原句】　本规则有几个例外。

【英语错句】　There is a few exceptions for this rule.

【错误分析】　（1）"a few"是用以修饰可数名词复数的，所以本句的主语是复数形式，因此应把"is"改成"are"。（2）在"exception"这个词后面只能用"to"，这属于固定的词汇搭配。

【改正后的句子】　**There are a few exceptions to this rule.**

**例 144**

【汉语原句】　应该指出，这种方法与现有的方法相比有许多优点，例如取样网络比较简单、成本低，等等。

【英语错句】　It should point out that this method has many advantages compared with those available, such as simple in sample network, and low in cost, etc.

【错误分析】　（1）本句主句部分并没有说明动作执行者，所以要用被动语态，改成"It should be pointed out"。（2）在名词"advantage"之后表示"与……相比"之意时一般应该用"over"，这是固定搭配。（3）由于"such as"引出的同位语是说明"优点（advantages）"的，所以应该使用名词短语而不能用形容词短语，应把"simple in sample network"改成"a simple sample network"，而把"low in cost"改成"low cost"。（4）由于用了"etc."，所以在第二个优点前不能用"and"，因为"and"表示结束了。

【改正后的句子】　**It should be pointed out that this method has many advantages over those available, such as a simple sample network, low cost , etc.**

**例 145**

【汉语原句】　根据上面对该多项式分解的分析，就得到了一种新颖的排列。

【英语错句】　On the basis of the analysis above to the decomposition of the polynomial, a novel kind of configuration is followed.

【错误分析】　(1)"对……的分析"实际上属于一种逻辑上的动宾关系,所以这里的介词"to"应该换成"of"。在英语中,由及物动词演变来的名词的逻辑宾语绝大多数应该由介词"of"来引出。(2)考虑到前后词之间关系的紧密,由于"above"放在被修饰词前后时含义不变,因此这里最好把"above"放在"analysis"前,这样"analysis"与其逻辑上的宾语就靠近了。(3)"kind of"在此是不该使用的,与"这种方法"的表示法一样,这里汉语中的"种"字是不能译出来的。(4)如果要表示"得到了"可以用"…is obtained"而不能用"is followed"。更好的表示方法可以使用不及物动词"results",这是英美人惯用的简洁表示法。

【改正后的句子】　**On the basis of the above analysis of the decomposition of the polynomial, a novel configuration results.**

例 146

【汉语原句】　在 1.2 节,我们将主要给出计算核和核度的一种递归公式。

【英语错句】　In the section 1.2, we shall mainly give a kind of recursive formula of computing the core and coritivity.

【错误分析】　(1)表示"1.2 节"时,在"section"前不得有冠词,且其第一个字母要大写,即"Section 1.2"。(2)在表示"一种公式"时不该使用"kind of"。(3)表示用于什么目的的公式(formula),算法(algorithm),语言(language)等名词后应该使用"for"而不用"of"。

【改正后的句子】　**In Section 1.2, we shall mainly present a recursive formula for computing the core and coritivity.**

例 147

【汉语原句】　证明了原问题的弱解的存在。

【英语错句】　The existing of weak solution of original problem is proved.

【错误分析】　(1)表示"存在"的名词为"existence",不该使用动名词"existing"。(2)在可数名词"weak solution"及"original problem"前均应加一个冠词,在此应该用定冠词"the",因为它们在此都是特指。(3)在名词"solution"后最常见的情况是:若该名词表示解的结果,则其后面一般用"to"(也有用"for"的);若该名词表示解这个动作,则应该用"of"了。

【改正后的句子】　**The existence of the weak solution to the original problem is proved.**

**例 148**

【汉语原句】　这电容器上的电流为零点几安培。

【英语错句】　The current on this capacitor is zero point several ampere.

【错误分析】　这是一个典型的中文式英语句。(1) 由于电流是流过元件的,常见的用法是在元件前用"through"而不是"on"(也可以用"in")。(2) 原作者不了解"零点几""零点零几""零点零零几"等的表达方法。在前面的例句中曾解释过,英美科技人员使用以下这一公式:

**a few [several] tenths [hundredths;thousandths]**

当后面跟有电压等单位时,应该使用"of a[an]+单位"。所以本句的表语应该为"a few tenths of an ampere"。又如:"这个电阻上的电压为零点零几伏特。"应该表示为:The voltage across this resistor is **a few hundredths of a volt.**

【改正后的句子】　**The current through[in] this capacitor is a few tenths of an ampere.**

**例 149**

【汉语原句】　我们设计了双星工作在几个不同的频率上,从而克服了盲速问题。

【英语错句】　We have designed dual satellite to work on several different operating frequencies, then the problem of blind speed is overcome.

【错误分析】　(1) 尽管汉语中用了"设计了",但根据其实际含义,应该把它改成"使得(make)"。(2) 在 "双星"前面应该加定冠词。(3) 在"frequency"前应该使用介词"at"。(4) 从语法上讲,在"then"之前要加上并列连词"and",不过这里最好把第二个句子改成分词短语作结果状语,使前后关系紧凑,即:thus overcoming the problem of blind speed。

【改正后的句子】　**We have made the dual satellite work at several different operating frequencies, thus overcoming the problem of blind speed.**

**例 150**

【汉语原句】　本文提出了估算频率偏移的一种新方法,其精度高,计算量少。

【英语错句】　A new kind of technique to estimate frequency deviation is proposed with high accuracy and low computation complexity.

【错误分析】　(1)"一种新方法"为"a new technique"而应该去掉"kind of"。(2) 在方法"method"或"technique"后最好用"for doing...",表示目的。

(3) 句中用了"with"短语，从语法讲它是修饰谓语动词"propose"的，这显然违背了汉语原句想要表达的含义，因此应该把它改成一个定语从句，写成"which gives a high accuracy and requires a small amount of computation"，这样就是修饰"a new technique"的了。

【改正后的句子】 **A new technique for estimating frequency deviation is proposed which gives a high accuracy and requires a small amount of computation.**

### 例 151

【汉语原句】 这样，天线场地的面积可以减少到通常的三分之一。

【英语错句】 In this way, the area of antenna site may be reduced as small as one-third of the conventional one.

【错误分析】 (1) 在"antenna site"前应加有不定冠词"an"。(2) "减少到"的"到"字的含义没有表达出来，应该在"reduced"之后加上介词"to"。(3) 去掉"as small as"，而要在"one-third"之后加代词"that"，它代替了名词"area"，而"one-third"是修饰"that"的，分数词和倍数可以直接修饰其后面的名词(也可以在其后面加"of")。

【改正后的句子】 **In this way, the area of an antenna site may be reduced to one-third that of the conventional one.**

### 例 152

【汉语原句】 他们并没有解释潜在的消费者如何能够获得在广告中出现的比喻所要达到的含意。

【英语错句】 They give no explanation on how potential consumers can get the intended meaning of metaphor appeared in advertisement.

【错误分析】 (1) "give"最好使用过去时"gave"。(2) 在"explanation"后应该使用"of"来引出它的逻辑宾语而不能用"on"。(3) "含义"是"implication"一词。(4) 在"metaphor"和"advertisement"前应该有冠词。(5) 由于"appear"是不及物动词，因此要用其现在分词作定语，而不能使用过去分词。

【改正后的句子】 **They gave no explanation of how the potential consumers can get the intended implications of the metaphor appearing in an advertisement.**

### 例 153

【汉语原句】 给出了由一种定义导出另一种定义的解析表达式。

【英语错句】　The analytic expression from one definition to derive the other is given out.

【错误分析】　（1）在"expression"之后一般跟"for …"，在此应该写成"for deriving one definition from the other"。（2）"given out"中的"out"是多余的，这是由于受到汉语的影响。

【改正后的句子】　**The analytic expression for deriving one definition from the other is given.**


例 154

【汉语原句】　本文研究了一种积分约束的波束形成器，以克服多点约束的波束不仅占用较多的权矢量自由度，而且抑止干扰的性能不稳健。

【英语错句】　This paper researches an integral constraint beamformer to overcome the shortcoming in multi-point constraint beam-forming that requires a large number of the degree of freedom in the weight vector and is not robust to the suppression of interference.

【错误分析】　（1）根据中文含义，句中的"researches"应该改成"presents"或"introduces"为好。（2）"beamformer"应该改成"beam-shaper"；在它后面一般要用"for …"为好。（3）由汉语得知，这里提到了两个缺点，所以"shortcoming"应该为复数形式。其后应该跟有两个分别由"that"引导的同位语从句，而在从句中要加上主语"it"。（4）"较多的"不能用"a large number of"表示，而且这个词组只能接可数名词复数，而这里是"the degree of freedom"，如果用"a lot of"又过分了，所以改成"much of"为好。

【改正后的句子】　**This paper presents an integral constraint beam-shaper for overcoming the shortcomings in multi-point constraint beam-forming that it requires much of the degree of freedom in the weight vector and that it is not robust to the suppression of interference.**


例 155

【汉语原句】　这表明期望收入严格按补偿速率增加，而当投标人足够多时最大的期望收入几乎没有差别。

【英语错句】　This shows that the expected revenues strictly increase in the compensating rate and when bidders are enough the maximum expected revenues are almostly no difference.

【错误分析】　（1）"strictly"在此应该处于"increase"之后，因为它是修饰

"按补偿速率"的。(2)在"速率"前应该使用介词"at"。(3)本句中主语从句的内容有两个,所以应该使用两个主语从句引导词"that",即在"when"之前要加上"that"。(4)"几乎"应该是"almost",并不存在"mostly"一词。(5)"投标人足够多"应该写成"the number of bidders is large enough"。(6)"没有差别"在此应该写成"remain the same"。

**【改正后的句子】** **This shows that the expected revenues increase strictly at the compensating rate and that when the number of bidders is large enough,the maximum expected revenues remain almost the same.**

**例 156**

**【汉语原句】** 在光纤承载射频系统中,传统单边带调制产生的光载波分量和光边带分量之间功率差较大。

**【英语错句】** In the radio over fiber(RoF)system,the difference of the power of optical carrier and that of optical sideband is large using the conventional optical single sideband with carrier(OSSB+C)modulation.

**【错误分析】** (1)本句应该采用英美人常用的句型"the difference in … between A and B"。(2)在"using"前加"when"。

**【改正后的句子】** **In the radio over fiber(RoF)system,the difference in power between optical carrier and optical sideband is large when using the conventional optical single sideband with carrier(OSSB+C)modulation.**

**例 157**

**【汉语原句】** 该算法引入了自适应 SINR 门限以及"投票模型",充分考虑了 LTE 网络中的小区类型。

**【英语错句】** The SINR threshold and "vote model" are introduced in this algorithm. It fully takes into account the cell type in the LTE network.

**【错误分析】** (1)最好把"In this algorithm"放在句首。(2)应该把两个句子合并成一个句子,句号变为逗号,"It"改为"which"。(3)"fully"改为"full"并放在"account"之前,又如:make **full** use of …; pay **much** attention to…。

**【改正后的句子】** **In this algorithm,the SINR threshold and "vote model" are introduced,which takes into full account the cell type in the LTE network.**

**例 158**

**【汉语原句】** 考虑到条件(4)、(6),我们可以得到下面的表达式。

**【英语错句】** Considered the condition（4），（6），we can obtain following expression.

**【错误分析】** （1）这里的"考虑到"的实际含义是"根据"，所以应该使用"by 短语"。（2）由于有两个条件，所以名词"condition"应该用复数形式，这是有些读者经常出错的，由于"条件"后面有号码，因此其前面一般不用冠词了。（3）凡有两个东西并列时，在科技写作中要在它们之间加上"and"一词，如果有多个东西并列，则要在最后两个东西之间加上"and"。（4）由于"expression"是可数名词单数且是确定的，所以应该在"following"之前加定冠词。

**【改正后的句子】** **By conditions（4）and（6），we can obtain the following expression.**

**例 159**

**【汉语原句】** 这个准则是最小均方准则(LMS)的一种高阶改进形式。

**【英语错句】** This criterion is a higher-order form improved from the Least Mean Square（LMS）criterion.

**【错误分析】** （1）"improved"应该放在"higher-order"之前，所以不定冠词应该用"an"。（2）"from"应该改成"of"。

**【改正后的句子】** **This criterion is an improved higher-order form of the Least Mean Square（LMS）criterion.**

**例 160**

**【汉语原句】** 强调指出了光强随圆柱体直径变化的正弦规律。

**【英语错句】** The sine law of the variance of the light intensity brought due to the variance of the cylinder diameter has been pointed out emphatically.

**【错误分析】** （1）在科技文中表示"变化"的名词常用"variation"，也可以用"change"，而很少使用"variance"的，尽管其词义与上述两词类同。（2）"A 随 B 的变化"属于一种固定的搭配模式，应该使用"the variation of A with B"。（3）"point out emphatically"这一概念虽然没有错，但它常用在普通英语和政治性文章中，在科技文中常用动词"emphasize"一词。

**【改正后的句子】** **The sine law of the variation of（the）light intensity with（the）cylinder diameter has been emphasized.**（根据观察，在搭配模式"名词＋of＋A＋介词＋B"时，名词"A"和"B"前的定冠词英美科技人员常常省去。）

**例 161**

【汉语原句】　在第一节对模拟计算和数字计算作了详细的比较。在这一节我们将讨论其他两类特殊的问题。

【英语错句】　In the section 1，the detailed comparison of analog and digital computing is given. In this section，we'll discuss other two types of special problems.

【错误分析】　(1) 在"section"前不能加冠词，而且其第一个字母要大写。(2) "detailed"之前的"the"一般应该改为"a"。(3) 与"comparison"搭配的动词常见的是"make"，我们会经常见到"make a comparison""make an analysis""make a study""give an introduction""give an explanation""give a description"等等，希望读者在阅读英语书刊时多留意不同的搭配关系。(4) 第一个句子的谓语使用一般过去时，因为现在开始讲第二节了，那么第一节就代表已经过去的事了(但也有个别人使用现在完成时)。(5) 由于科技文是一种严肃的文体，所以一般不使用紧缩表示法，因此"we'll"应该写成"we shall [will]"，同样"can't"应写成"cannot[can not]"，"it's"应写成"it is[it has]"等。(6) 英语的习惯是要把"other"放在表示数量的词之后，即应该写成"two other"，这是在少数情况下英汉的词序是不同的，又如：**each receiver** section(接收机的每一部分)""**four additional[added]** clock pulses(另外四个时钟脉冲)""**advanced world** level(世界先进水平)""**real/actual current** direction(电流的实际方向)""**maximum instantaneous transistor** current(晶体管的最大瞬时电流)""a **major theoretical** advance(理论上的一个主要进展)""**a few short** years(短短几年)""**iron and steel** industries(钢铁工业)""**plants and animals**(动植物)""**northwest**(西北)"等。(7) "special"应该放在"type""form""sort""kind"等之前，又如："a **new** pair of shoes(一双新鞋)""a **new** set of parameters(一套新的参数)"等。(8) 在"type[sort；kind；form] of"之后一般用不带冠词的单数名词。

【改正后的句子】　**In Section 1, a detailed comparison of analog and (或：with) digital computing (或：A detailed comparison between analog and digital computing) was made. In this section we shall discuss two other special types of problem.**

**例 162**

【汉语原句】　本文分析了不同表面均方根斜率对遮蔽效应的影响。

【英语错句】　This paper analyzes the influence of the shadowing effect for different surface RMS slope.

【错误分析】　(1) 本句的主要错误是原作者不了解词汇搭配问题，即"the effect [influence；impact] of A on[upon] B(A 对 B 的影响)"。(2) "slope"在此应该用复数形式。

【改正后的句子】　**This paper analyzes the influence of different surface RMS slopes on the shadowing effect.**

例 163

【汉语原句】　这新算法能给出类似于经典算法的 SNR 值。

【英语错句】　This new algorithm can give the similar SNR as the classic algorithm.

【错误分析】　(1) "similar"应该与介词"to"连用。(2) "the"应该改成不定冠词。(3) "经典算法的"在此的实际含意是"that (given) by the classic algorithm"，否则比较对象就不一致了。

【改正后的句子】　**This new algorithm can give an SNR similar to that by the classic algorithm.**

例 164

【汉语原句】　导体的电阻不仅取决于制成导体的材料，而且也取决于导体的尺寸和温度。

【英语错句】　The resistance of a conductor not only depends on the material with which the conductor is made，but also on the size and temperature of the conductor.

【错误分析】　(1) "由……制成"应该是"be made of"，所以定语从句中"with which"应改成"of which"。(2) "not only … but also …"属于等立连词，其连接的两个东西应该是平行的，本句中"not only"后跟动词"depends"而"but also"后跟介词"on"，为了一致起见，"not only"应该放在"depends"之后。(3) 在修饰"不及物动词＋介词"型短语动词中副词或介词短语的位置往往应放在短语之中的介词之前。例如：

这台计算机主要由五部分组成。

This computer consists **mainly** of five units.

这是部分地由于地球的自转引起的。

This is due **in part** to the earth's rotation.

【改正后的句子】 **The resistance of a conductor depends not only on the material of which the conductor is made, but also on the size and temperature of the conductor.**


### 例 165

【汉语原句】 在每一给定的间隔内可自动地计算出完成的工作量与计划的工作量之比。

【英语错句】 In every given interval the ratio of the finished work amount and the scheduled work amount is automatically calculated.

【错误分析】 本句主要错误在于作者不熟悉 ratio 这个词的搭配关系，它有两种搭配关系：第一，"the ratio of A to B"；第二，"the ratio between A and B"，它们都表示"A 与 B 之比"，但以前者为多见。

【改正后的句子】 **In every given interval, the ratio of the finished work amount/load to the scheduled one is automatically calculated.**


### 例 166

【汉语原句】 在这个方法中，把这两个输入矢量并合成一个块矢量，它可以表示如下。

【英语错句】 In this method, the two input vectors are grouped to be a block vector and which can be expressed as following.

【错误分析】 (1) 动词"group"是不能与"to be …"搭配使用的，它只能与"as …"或"into …"搭配。(2) 由于"which"是引导定语从句的，所以在其前面是不能有"and"的，因为这两部分并不并列。(3) "as follows"是固定词组，而不能写成"as following"，有的读者不时地会犯这个错误。

【改正后的句子】 **In this method, the two input vectors are grouped as a block vector which can be expressed as follows.**


### 例 167

【汉语原句】 最后用计算机仿真对窗口损耗概率进行了分析。

【英语错句】 In the end, analysis of packet loss probability is presented by computer simulation.

【错误分析】 (1) 表示"最后"应该用副词"finally"。(2) "分析""研究""比较"等之前一般要用不定冠词。（3）"进行"一般使用动词"make"（也有用"conduct"的）。

【改正后的句子】　**Finally，an analysis of packet loss probability is made/conducted by computer simulation.**

**例 168**

【汉语原句】　本文介绍了用于估算多普勒调频率的最小熵方法，它与经典方法相比具有精度高、计算量小的优点。

【英语错句】　The minimum entropy technique to estimate the Doppler frequency rate has been presented. Its high accuracy and small amount of computation is advantageous over the classical technique.

【错误分析】　（1）在"technique 后通常跟"for …"。（2）本句主要错误在于比较的对象不一致。（3）汉语的一整句话写成了英语两句话，显得很松散，所以应该把英文的两句合并成一句，把第二句改成一个非限制性定语从句。

【改正后的句子】　**The minimum entropy technique for estimating the Doppler frequency rate has been presented，which has the advantages over the classical techniques of high accuracy and a small amount of computation.**（这里"of"引出了"advantages"的同位语。）

**例 169**

【汉语原句】　当信噪比低的时候，噪声的平均值并不是完全为零，而其值大约具有与信号相同的量级。

【英语错句】　When the SNR is low, the mean of the noise is not perfect zero and the value of which is about the same order with the signal.

【错误分析】　（1）"perfect"应该用其副词形式，即"perfectly"。（2）后面的定语从句是错误的，它不应该与并列连词"and"连用。为了使句子紧凑而精炼，可以把"and …"这一部分用一个"with 结构"来表示。（3）"the same"必须与"as"搭配使用。

【改正后的句子】　**When the SNR is low，the mean of noise is not perfectly zero，with its value being（of）about the same order as the signal.**

**例 170**

【汉语原句】　利用这种传感器完成了高温测量，其测量精度好于 0.1%。

【英语错句】　Using the sensor of this kind, the high-temperature measurement has been achieved. The precision of the measurement is better than 0.1%.

【错误分析】　(1) 分词"using"的逻辑主语并不是"the high-temperature measurement"，虽然看得懂，但最好用"By using…"或"By (the) use of…"。(2) 与"measurement"相搭配的动词是"make"而不是"achieve"。(3) 汉语原句中后一部分是对前一部分的附加说明，两者应紧密结合在一起，不应该单独译成一句，所以这里我们可以用一个"with 短语"来表示，写成"with a precision of…"。

【改正后的句子】　**By use of the sensor of this kind, the high temperature measurement has been made, with a precision of better than 0.1%.**

例 171

【汉语原句】　电子货币与物理货币相比具有多方面的优势，但其重用性未获得充分重视。

【英语错句】　Compared with physical cash, electronic ones possess much more advantages, but reusability is paid less attention.

【错误分析】　(1) 凡是见到"advantage"时，"与……相比"的含义就应该使用"over"来表示，这样句中就不会出现比较级的形式。(2) "pay attention to…"的被动形式应该是"… is paid attention to"或"Attention is paid to …"。(3) "cash"为不可数名词，所以不能使用代词"ones"。

【改正后的句子】　**Electronic cash possesses many advantages over physical cash, but its reusability is not paid full attention to.**

例 172

【汉语原句】　这个系数具有与公式(1)中相似的含义。

【英语错句】　This coefficient has similar meaning as in the formula (1).

【错误分析】　(1) 由于"meaning"是可数名词单数，所以其前面应该有冠词。(2) "similar"是与介词"to"连用的（"the same"才与"as"连用）。(3) 在"in"之前应该加上代词"that"，否则比较对象不一致。(4) "the formula (1)"应该改成"Formula (1)"，即其前面不得有冠词，而且其第一个字母要大写。

【改正后的句子】　**This coefficient has a similar meaning to that in Formula (1).**

### 例 173

【汉语原句】 所有这一切为设计和研究高温 6H‑SiC CMOS 电路提供了有益的参考。

【英语错句】 All these provide valuable reference to the design and research of high temperature 6H-SiC CMOS circuits.

【错误分析】 （1）表示"所有这一切"的"All these"应该写成"All this"，这是英美人的习惯，它作主语时谓语要用单数形式。（2）"provide"的用法是"provide sb. **with** sth."或"provide sth. **for** sb."，所以本句中的"to"应该改为"for"。（3）"reference"在此是一个可数名词，所以其前面要有冠词。（4）由于"design"和"research"后跟的逻辑宾语要用不同的介词引出，所以这里不能用"of"来统管这两个名词，而是应该表示为"the design **of** and the research **on**"。

【改正后的句子】 **All this provides a valuable reference for the design of and the research on high temperature 6H‑SiC CMOS circuits.**

### 例 174

【汉语原句】 每个 eBOX 安装了 WINCE 操作系统。

【英语错句】 Each eBOX is installed WINCE operating system.

【错误分析】 （1）本句主要错在如何表示"安装了"，实际上这里的"安装了"的真实含义是"装[配]备了"，所以应该使用"be equipped with …"。（2）由于这里的"WINCE 操作系统"是个特定的可数名词单数，所以在其前面应该有一个定冠词。

【改正后的句子】 **Each eBOX is equipped with the WINCE operating system.**

### 例 175

【汉语原句】 本文分析了影响通道均衡的几个因素，并给出了通道均衡的方法。

【英语错句】 This paper analyzes several factors that affect on channel equalization, and gives the method of channel equalization.

【错误分析】 （1）在"factors"后，英美人一般使用"affecting"而不必用定语从句的。（2）由于"affect"是及物动词，所以其后面不能加介词"on"。（3）在"method"后多数人用"for …"。

【改正后的句子】 **This paper analyzes several factors affecting channel equalization, and gives the method for channel equalization.**

**例 176**

【汉语原句】　随着时间的推移，人们会习惯这一颜色的。

【英语错句】　People will get used with this color as time flies.

【错误分析】　(1)"get used[accustomed]"要与"to"连用。(2)"随着时间的推移"应该表示为"as time goes on"或"with the lapse of time"。

【改正后的句子】　**People will get used to this color as time goes on.**

**例 177**

【汉语原句】　这方法适合于分析任意给定形状的复杂结构。

【英语错句】　This method is suitable to analyze complex structure of arbitrary given shape.

【错误分析】　(1)"suitable"后面跟的"to"是介词而不是不定式的标志，不少人喜欢用介词"for"。(2)在"complex structure"之前应该加定冠词。(3)在"of"之后应该加不定冠词"an"。(4)形容词"arbitrary"应该改为副词"arbitrarily"，因为它是修饰分词"given"的。

【改正后的句子】　**This method is suitable for analyzing the complex structure of an arbitrarily given shape.**

# Ⅳ、代词

**例 178**

【汉语原句】　每个子问题能够用其所有周围子问题的解来解。

【英语错句】　Each subproblem can be solved with the solution of its all surrounding subproblems.

【错误分析】　(1)"solution"应该用复数。(2)"solution"后跟"to"或"for"。(3)"its all"要改成"all its"，"all"一定要处于一个短语的最前面。

【改正后的句子】　**Each subproblem can be solved with the solutions to all its surrounding subproblems.**

**例 179**

【汉语原句】　开关电源(SMPS)输入端的电磁干扰(EMI)频谱分析及其抑制

【英语错句】　EMI spectrum analysis and restraint in input of SMPS

**【错误分析】** 这是一篇论文的标题。(1)"电磁干扰频谱分析"的含义是"对电磁干扰频谱作一分析",所以应该写成"an analysis of the EMI spectrum"。(2)"在输入端"应写成"at the input"。(3)由于"开关电源"是一个可数名词,所以在"SMPS"前应该加一个定冠词"the"。(4)"其抑制"意为"对它进行抑制",一般应该使用"its suppression"表示,这里"its"与"suppression"之间的逻辑关系为"动宾关系",也就是说,"its"在逻辑上是"suppression"的对象。根据观察,在英语中,物主代词作定语时可能存在以下四种逻辑关系(了解了这些逻辑关系对理解原文是非常有帮助的,而且写出来的句子很简练,属于地道的科技英语文体):

(1)"所属关系"

这时,把物主代词译成"……的"。例如:

**Its** advantages are small size and low cost.

它的优点是体积小、成本低。

(2)"主表关系"

这时物主代词所修饰的名词来自形容词,所以相当于主语与表语之间的关系,一般要把物主代词译成人称代词。例如:

Water and carbon dioxide are among substances that absorbs in the infrared. **Their** presence in the atmosphere has an insulating effect.

水和二氧化碳属于能够吸收红外线的物质。它们存在于大气中[它们在大气中的存在]起到了隔离作用。

(3)"主谓关系"

这时一般应该把物主代词译成人称代词。

① 物主代词所修饰的名词来自于不及物动词的情况

A baseball that lands in an open soon comes to rest because of **its** interaction with the ground.

落在开阔地上的棒球,由于它与地面的相互作用很快就会停下来。

——"its"是"interaction"的逻辑主语;"because of **its** interaction with the ground"等效于"because it interacts with the ground"

② 物主代词所修饰的名词来自于及物动词的情况

Thus far **our** discussion of the principles of mechanics has been concerned primarily with particles.

到目前为止,我们对力学原则的讨论主要涉及了质点。——"our"是"discussion"的逻辑主语;而"of the principles of mechanics"是"discussion"的逻辑宾语

This is possible because of **our** assumption that the transistor is a linear amplifier over the range of voltages and currents of interest.

之所以能这样是因为我们假设晶体管在我们感兴趣的电压、电流范围内是一个线性放大器。——"our"是"assumption"的逻辑主语，而"that …"是它的逻辑宾语从句，纯语法上是它的同位语从句

（4）"动宾关系"

有时可看成"被动的主谓关系"——这时物主代词所修饰的名词来自于及物动词（物主代词要译成人称代词）。

The results of an experiment forced the abandonment of Thomson's model and **its** replacement by a new model.

一个实验的结果迫使人们放弃了汤姆森模型，而用一个新的模型来取代它。——"its"是"replacement"的逻辑宾语；也可等效成"**it** was replaced by a new model"

Attention focuses on the floated gyroscopes because of **their** utilization as an essential component in ballistic missiles.

我们侧重于讲浮型陀螺仪，因为它们被用作为弹道导弹中的一个关键部件。——我们既可以把"their"看成是"utilization"的逻辑宾语，也可把它看成是"utilization"的被动的逻辑主语，即"because of their utilization as an essential component in ballistic missiles"可以等效成"because they are utilized as an essential component in ballistic missiles"

**【改正后的句子】** **Analysis of the EMI spectrum at the input of the SMPS and its suppression**

**例 180**

**【汉语原句】** 在这种情况下，$r(t)$的变化是如此之小以至于可以忽略不计。

**【英语错句】** In this case, variations in $r(t)$ are so small that can be neglected.

**【错误分析】** 本句错在"以至于可以忽略不计"的表达上。由于汉语原句中省去了"忽略不计"的对象，因而造成了引导结果状语的 that 从句中缺少了主语。因为主句中作主语的名词"variations"为复数形式，所以在"that"之后应该加人称代词"they"。当然也可以用"so…as to (do)"这一短语来表达，也就是把"that can"改写成"as to"就行了。

**【改正后的句子】** **In this case，variations in $r(t)$ are so small that they can**

be neglected（或：…so small as to be neglected）.

例 181

【汉语原句】　这两位工程师正忙于设计一种新型的软件。

【英语错句】　This two engineers are busy to design a type of new software.

【错误分析】　（1）这里的"这"在此指"两位"，所以根据语法应该用"These"而不用"This"。（2）在形容词"busy"之后按传统规则有两种接续方式，一种是"with the design of …"，另一种是"(in) designing …"，两种都可以。（3）"new"的位置错了，应该放在"type"之前。

【改正后的句子】　These two engineers are busy with the design of（或：busy (in) designing）a new type of software.

例 182

【汉语原句】　格的最主要特点是将矢量空间 $R^n$ 进行划分，这在矢量量化和编码理论中具有重要的现实意义。

【英语错句】　The most important characteristic of lattices is that it divides the vector space $R^n$, and thus there is an important practical significance in the application of lattices to the vector quantization and coding theory.

【错误分析】　（1）"lattices"应该用单数形式。（2）"具有意义[重要性、价值、帮助等等]"应该用"be of＋这些抽象名词"来表示，有时也可以用"has"或"have"，不能用"there be"句型，同时应该用零冠词（即不用任何冠词）。（3）为了使英语句子显得紧凑，把逗号后的那部分用非限制定语从句来表示比较好。

【改正后的句子】　The most important characteristic of the lattice is that it divides the vector space $R^n$, which is of important practical significance in the application of the lattice to the vector quantization and coding theory.

例 183

【汉语原句】　这三个柱矢量中的任何一个都可以被表示成其他两个的线性结合。

【英语错句】　Any of this three column vectors can all be expressed as the linear combination of two other.

【错误分析】　（1）"this"应该改写成"these"。（2）"all"应该去掉，这两个错误都是受到了汉语的影响。(3)"其他两个"可有两种方式来表示："the other two"或"the two others"。

**【改正后的句子】　Any of these three column vectors can be expressed as the linear combination of the other two.**

例 184

【汉语原句】　提出了一种基于模糊熵的多值图像恢复方法,该方法根据熵在应用方面的局限性,对其表示的形式进行了扩展,构造了一类能够反映多值图像特点的模糊熵。

【英语错句】　A method of multi-level image restoration based on fuzzy entropy is presented. The method can extend the entropy's express form in terms of entropy's limitations to application and construct a kind of fuzzy entropy that can reflect characteristic of multi-level image.

【错误分析】　(1)在"method"后,英美人喜欢用介词"for"。(2)为了使句子紧凑,原来的两个句子应该合并成一句,把第二句改写成一个非限制性定语从句,即："… is presented,which can …"。(3)"express form"应该改成"expression"。(4)在"limitation"后常跟"on"。(5)在"application"前应该加上"its"。(6)"characteristic"应该用复数形式。(7)在最后的"multi-level image"前应该加一个冠词。

**【改正后的句子】　A method for multi-level image restoration based on fuzzy entropy is presented,which can (be used to) extend the entropy's expression in terms of entropy's limitations on its application and construct a kind of fuzzy entropy that can reflect characteristics of the multi-level image.**

例 185

【汉语原句】　其目的、前提、解题方法是不同于其他模型的。

【英语错句】　Its objective、premise、to solve the problem differ from that of other models.

【错误分析】　(1)英语中是没有顿号的,应该使用逗号来表示,而且在最后两个之间应该加上"and"。(2)英语中,"solution"表示"解"或"解题方法"两个意思,它都是使用"solution to the problem"来表示,这样就与前面两个名词并列了。(3)由于主句的主语是三个东西,因此,应该把比较从句中的"that"改为"those"。

【改正后的句子】 **Its objective, premise, and solution to the problem differ from those of other models.**

**例 186**

【汉语原句】 同经典的 IMM－EKF 算法相比，这两种算法的鲁棒性和精度都是优越的。

【英语错句】 The robust and accuracy of these two algorithms are more advantage over that of the classical IMM-EKF algorithm.

【错误分析】 (1) 鲁棒性应该用名词"robustness"来表示。(2) "比……优越"应该用"superior to"表示。(3) "that"应该改成"those"，因为它代替了两个东西。

【改正后的句子】 **The robustness and accuracy of these two algorithms are superior to those of the classical IMM-EKF algorithm.**（更好的版本为：**These two algorithms are superior in robustness and accuracy to the classical IMM-EKF algorithm.**）

# V、"and"的使用

**例 187**

【汉语原句】 首先，从数据点到质心的距离对迭代过程进行约束。

【英语错句】 Firstly, the iteration process is constraint from the distance between data point to centroid.

【错误分析】 (1) "Firstly"意为"第一"，是用来列举条目的。这也是不少读者写作时常常犯的错误，同样"secondly"要改成"second"，"thirdly"要改成"third"等，因此这里要改成"First"。(2) "constraint"是名词，这里应该用动词"constrain"的过去分词。(3) 应该把"to"改成"and"，因为是"between A and B"，等同于"from A to B"。

【改正后的句子】 **First, the iteration process is constrained from the distance between data point to centroid.**

**例 188**

【汉语原句】 这里 $a_i, b_i, c_j, d_j$ 是标量，其中 $i=1, 2$ 而 $j=1, 2, 3$。

【英语错句】 Here $a_i$, $b_i$, $c_j$, $d_j$, $i = 1, 2$, $j = 1, 2, 3$, are scalars.

【错误分析】 (1) 英语中若有几个东西并列的话，一般在最后两个之间应该加一个并列连词"and"，也就是"A and B""A, B(, ) and C""A, B, C(, ) and D"等等，英国人一般不用括号内的那个逗号，而美国人喜欢用；如果特别强调的话，可以在每两个之间加上一个"and"。有多个并列的同类从句时，"and"也是这样用的。所以在"$c_j$"和"$d_j$"之间应该加一个"and"；同样在"$j$"之前也要加"and"。(2) "$i$"和"$j$"是对前面参量的附加说明，所以一般使用"with结构"为好，也可用"where"引出的定语从句。(3) 原来的英语句中第 4、6、9 个逗号应该去掉。

【改正后的句子】 Here $a_i$, $b_i$, $c_j$, and $d_j$ with $i = 1, 2$ and $j = 1, 2, 3$ are scalars.

例 189

【汉语原句】 通过理论分析和实验证明该方法是一种估计 SVM 的推广性能并且可把它应用于范围很广的模式识别问题的一般方法。

【英语错句】 This method is proved to be a general method for estimating the generalization performance of a SVM by theoretic analysis and experiments and can be applied to wide range problems of pattern recognition.

【错误分析】 (1) 对于本句的主干部分英美人常用的是主动形式，也就是"Theoretical analysis and experiment(s) show that …"这样显得比较简练。(2) 由于使用了主动形式后，在"show"后面就有两件事，这时英美人喜欢用两个由"that"引导的宾语从句，在第二个"that"前要加"and"。又如：

Some physical quantities require only a magnitude and a unit to be completely specified. Thus it is sufficient to say **that** the mass of a man is 85 kg, **that** the area of a farm is 160 acres, **that** the frequency of a sound wave is 660 cycles/sec, **and that** a light bulb consumes electric energy at the rate of 100 watts.

有些物理量只要明确标出大小和单位就可完全说明清楚了。所以只要说出以下这些就够了：一个人的质量为 85 千克；某农场的面积为 160 英亩；某声波的频率为 660 周/每秒；一个灯泡消耗的功率为 100 瓦。(在第二个长句中，"to say"后面跟了四个由"that"引导的宾语从句。)
(3) "范围很广的"应该写成"a wide range of"，而不能把"wide range"直接修饰到其后面的名词"problems"上去；可是为了简明起见，可以把"pattern recognition"直接放在"problems"的前面作前置定语，至于在什么情况下可以用一个名词短语作前置定语，希望读者在阅读科技文献时多观察，努力模仿外国人的用法。(4) 由于"SVM"是按字母读的,其第一个字母是元音,所以在其前面一定要用"an"而不是"a",这个小问题也是不少人经常出错的。

**【改正后的句子】** Theoretical analysis and experiment show that this method is a general method for estimating the generalization performance of an SVM and that it can be applied to a wide range of pattern recognition problems.

**例 190**

**【汉语原句】** 所谓"工程制图",我们指的是机械制图、地图等。

**【英语错句】** The so-called "engineering drawing", we mean a mechanical drawing, a geographic map, and etc..

**【错误分析】** (1) 名词短语"The so-called'engineering drawing'"在句中没有作用,成了一个多余的成分,所以这句子不对。关于"所谓 A(我们)指的是 B"应该使用固定句型"By A we mean B"或"By A is meant B"来表示。(2) 句中用了"etc."后其前面是不能使用"and"的。(3) 由于缩略词"etc."本身有一个黑点了,现在恰逢处于句尾,所以句号与这一点就重合了,有的读者往往在这点后再加一个句号,这是不对的。

**【改正后的句子】** By engineering drawing we mean a mechanical drawing, a geographic map, etc.

**例 191**

**【汉语原句】** $\delta$、$\delta_e$ 可以从式(3)和式(4)求得。

**【英语错句】** $\delta$、$\delta_e$ can be gotten from Eq. (3) and (4).

**【错误分析】** (1) 英语中是没有顿号的,只能用逗号来表示,可是这里只有两个东西,所以根据英语的规则在这两者之间必须用连接词"and"。(2) 在科技文的被动句中表示"得到"一般使用"obtain"而很少用"get"。(3) 由于这里有两个式子,所以要注意使用复数的名词"Eqs. (3) and (4)",又如"在图(3)、(4)中"应写成"in Figs. (3) and (4)","在第一、二章中"应写成"in Chaps. 1 and 2"等。不过要注意,若表示"从第 5 页到第 10 页"的话,则用"pp. 5~10"而绝不用"ps. 5~10"。

**【改正后的句子】** $\delta$ and $\delta_e$ can be obtained from Eqs. (3) and (4).

**例 192**

**【汉语原句】** 这种新方法具有效率高、容易调整的优点。

【英语错句】 This kind of new method have the advantage of high efficiency，easy adjustment.

【错误分析】 （1）"这种新方法"的表达法应该是"this new method"。（2）"have"应改成单数第三人称形式"has"。（3）由于汉语原句中提到了两个优点，所以"advantage"应该用复数形式。（4）由于本句只提及了两个优点，所以在两者之间应该使用 and 一词。

【改正后的句子】 **This new method has the advantages of high efficiency and easy adjustment.**

例 193

【汉语原句】 数字结果表明周期模式决定了中心频率，而圆孔半径影响着禁带的宽度和深度。

【英语错句】 Numerical results show that the period pattern determines the central frequency and the radius of holes influences the width and the depth of the forbidden band.

【错误分析】 由于动词"show"后跟两个概念，所以根据英语习惯应该在第一个"and"之后加上宾语从句引导词"that"；或者把第一个"and"换成从属连接词"while"也行，起到了前后对照的作用。

【改正后的句子】 **Numerical results show that the period pattern determines the central frequency and that the radius of holes influences the width and（the） depth of the forbidden band.**（括号中的"the"可以省去，这时与前面的名词共用了一个"the"。）

例 194

【汉语原句】 设 P 和 Q 为同阶的两个布尔排列，于是它们的合成是一个新的布尔排列。

【英语错句】 Let P and Q be two Boolean permutations of same order, then their composition is a new Boolean permutation.

【错误分析】 （1）在"same"前要有定冠词"the"，这一点往往被读者遗忘掉。（2）由于"let"表示的是一个并列分句，它并不是一个条件状语从句（尽管其表达的含义类同于一个假设条件；或者说："祈使句＋and"＝"if 条件句"），所以在"then"之前必须加上一个并列连词"and"，表示这两个句子是并列的（若一开头是一个条件状语从句，则"then"之前是不能加"and"的）。另一种修改方法是把逗号改成句号，然后由"then"开头另列一个句子，这时应写成"Then…"。

**【改正后的句子】** Let P and Q be two Boolean permutations of the same order，and then their composition is a new Boolean permutation.

**例 195**

**【汉语原句】** 光折变 MPPC 的响应时间可以大大缩短，因而这一技术对于低重复的窄脉冲射束的 MPPC 的应用具有巨大的实用价值。

**【英语错句】** Response time of photorefractive MPPC can be deduced greatly，hence this technique has great practical value for the application of MPPC with low repeated narrow pulse beam.

**【错误分析】** （1）在"response"、"photorefractive MPPC"和"MPPC"前应该加上定冠词。（2）"deduce"意为"推断"，应该把它改为"reduce"或"shorten"。（3）在"hence"前应该加上"and"。（4）"has"改为"is of"为好（它表示"具有"时后跟一些抽象名词，如"价值""用途""意义""重要性""帮助"等）。（5）"low repeated …"应该写成"a lowly repeated …"。

**【改正后的句子】** The response time of the photorefractive MPPC can be reduced greatly，and hence this technique is of great practical value for the application of the MPPC with a lowly repeated narrow pulse beam.

# Ⅵ、词序问题

**例 196**

**【汉语原句】** 我们所说的机器人就是一种特殊的电子设备。

**【英语错句】** What we call robot is just a kind of special electronic device.

**【错误分析】** （1）在可数名词单数"robot"前面应该有冠词。（2）"special"这个形容词的位置错了，应把它放在"kind"之前。类似的有"type""sort""set of …"等。

**【改正后的句子】** What we call a robot is just a special kind of electronic device.

**例 197**

**【汉语原句】** 这里的任何学生均不会解这类特殊的方程。

【英语错句】　Any student here cannot solve this type of special equation.

【错误分析】　（1）由于本句属于"全否定"的含义，所以英美人通常使用"No student here can"或"None of the students here can"（在"students"前一定要有定冠词，表示一群特定的人）这种表示方法。（2）"special"这个形容词的位置错了，应把它放在"type"之前。

【改正后的句子】　**No student（或：None of the students）here can solve this special type of equation.**

### 例 198

【汉语原句】　还有其他一些因素应该仔细加以考虑，例如政策、不可预见的自然灾害以及有关工厂的经济效益。

【英语错句】　There are also other several factors，such as policies，unpredictable natural disasters，and the economic efficiency of the factory involved，and should be carefully considered.

【错误分析】　（1）"其他一些"一定要使用"several other"，这是与汉语的词序不同之处，又如"其他两本书"应该写成"two other books"，"其他一些工厂"应该写成"some other factories"。（2）"经济效益"应表示"economic returns〔results；benefits〕"。（3）本句的句型应该采用"there be＋主语＋定语〔分词；形容词短语；不定式短语或定语从句〕"模式，在此"定语"部分本句最好用一个定语从句，写成"which should be carefully considered"。

【改正后的句子】　**There are also several other factors，such as policies，unpredictable natural disasters，and the economic benefits of the factory involved，which should be carefully considered.**

### 例 199

【汉语原句】　这六个月来他们一直在设计一种新型的计算机。

【英语错句】　They have been designed a type of new computer since these six months.

【错误分析】　（1）表示"一直在"应该使用"现在完成进行时"，所以应该把"designed"改成现在分词"designing"。（2）"new"这个形容词应该放在"type"之前。（3）"since"应该去掉，在"since"后应该是时间上的一点，而本句中表示持续的一段时间，当然在此也可用"for the last six months"来表示，而在由指示代词修饰的短语"these six months"前不用介词"for"。

【改正后的句子】 **They have been designing a new type of computer these six months.**

**例 200**

【汉语原句】 在这种情况下，二极管的反向电流可以忽略不计。

【英语错句】 In this case the diode reverse current is negligible.

【错误分析】 "二极管的反向电流"应该写成"the reverse diode current"，因为英语中最本质的词最靠近被修饰的名词。下面再举几例：the **forward diode** current（二极管的正向电流），the **positive battery** terminal（电池的正端），**common resistor** power ratings（电阻器通常的功率额定值），**preliminary circuit** designs（电路的预设计），the **principal remaining** limitation（余下的主要限制因素），**many flip-flop** applications（触发器的许多应用场合），**each following** stage（下面的每一级），the **possible Q-point** variation（Q 点可能发生的变化），**four additional ［added］** clock pulses（另外 4 个时钟脉冲），**two other** inputs（其他两个输入），the **outer atomic** electrons（原子的外层电子）。

【改正后的句子】 **In this case the reverse diode current is negligible.**

**例 201**

【汉语原句】 从结构级设计外部存储器的数据重组与访问方法，实现存储器的近似访问。

【英语错句】 The data reorganization and access method of external memory is designed from the structure level, and it can realize the access of approximate memory.

【错误分析】 （1）"method"后的介词"of"改成"for"为好。（2）在"external"前要加冠词。（3）把"from"改成"at"。（4）"structure"要改成其形容词"structural"。（5）形容词"approximate"应该放在"access of"前。（6）在句尾的"memory"前要加冠词。（7）为了使句子紧凑，把第二句改成一个非限制性定语从句。

【改正后的句子】 **The data reorganization and access method for the external memory is designed at the structural level, which can lead to the approximate access of the memory.**

**例 202**

【汉语原句】　每个中央处理装置（CPU）具有它知道如何执行的一套非常基本的功能。

【英语错句】　Each CPU has a set of very elementary function it knows how to perform.

【错误分析】　（1）形容词"very elementary"应该放在"set"前。（2）"function"一词应该用复数形式。

【改正后的句子】　**Each CPU has a very elementary set of functions it knows how to perform.**

# 第二部分　语 法 问 题

## Ⅰ、动词不定式

**例 203**

【汉语原句】　由于这个原因，这个算法需要进一步加以研究。

【英语错句】　Owing to this reason, this algorithm need study further.

【错误分析】　（1）应该把"owing to"改成"for"。（2）"need"应该用单数第三人称形式，即加"s"。（3）"study"应该改成"to be studied"或"studying"，这属于动词"need"的一种用法，这时其主语是"study"的对象。

【改正后的句子】　**For this reason, this algorithm needs to be studied（或：needs studying）further.**

**例 204**

【汉语原句】　小波变换非常适用于表征暂态现象的特征。

【英语错句】　The wavelet transform is very adapted to characterize transient phenomena.

【错误分析】　（1）"very"不能修饰动词的，要把它改成"well"。（2）"be adapted to（适合于）"这个短语中的"to"是个介词而不是动词不定式的标志，所以在此 to 之后应该使用动名词。这一点是不少人在写作时不怎么注意的，往往错把"to"当成动词不定式的标志。

【改正后的句子】　**The wavelet transform is well adapted to characterizing transient phenomena.**

**例 205**

【汉语原句】　这一技术能使传统检测应用于系统性能的评估。

【英语错句】　This technology can apply the traditional detection to the evaluation of system performance.

【错误分析】　本句属于中文式的英语，"technology"是不会发出"apply"

这一动作的，应该采用"make it possible to (do sth)"。

【改正后的句子】　**This technology makes it possible to apply the traditional detection to the evaluation of system performance.**

例 206

【汉语原句】　这种算法实现了代理人感知和作用于他们所处环境的能力。

【英语错句】　This kind of algorithm realizes the ability of agents in sensing and acting on the environment that they live.

【错误分析】　(1)"kind of"应该去掉。(2)由于"ability"的关系，所以"in sensing and acting"应该改写成"to sense and act"。(3)定语从句引导词"that"应该改成"where"或"in which"；若"that"不变的话，则也可以在从句末尾加上"in"，不过这种表示法在科技文中并不多见。

【改正后的句子】　**This algorithm realizes the ability of agents to sense and act on the environment where they live.**

例 207

【汉语原句】　这种材料很难加工。

【英语错句】　This material is very difficult to be machined.

【错误分析】　本句的错误是不少中国学生和科技人员常犯的，其关键是一定要用主动形式的动词不定式，即"to machine"。有人认为"材料"自己不会"加工"而是"被加工"的，所以要用被动式的动词不定式，主要是由于这些读者不了解"反射式不定式结构"的缘故。所谓"反射式不定式结构"可以被定义如下："句子的主语（或宾语）是句尾不定式（或不定式短语末尾的介词）的逻辑宾语"这样的一种结构形式。最常见的情况是句尾不定式的宾语占据了形式主语或形式宾语"it"的位置。例如：

It is not difficult to operate this computer.

操作这台计算机并不困难。

其反射式结构是"**This computer** is not difficult to operate."

这台计算机不难操作。

又如：

We find it difficult to explain the theory of relativity.

我们发现要解释相对论是很难的。

其反射式结构是"We find **the theory of relativity** difficult to explain."

我们发现相对论是很难解释的。

别的情况的例子有：

**This table** is unfit <u>for a student to make an experiment</u> <u>on</u>.

这张桌子不适合学生在上面做实验。（"This table"是介词"on"的逻辑宾语）

**This interior environment** is comfortable enough <u>for as many as four</u> <u>technicians to work</u> <u>inside</u>.

这种内部环境是相当舒适的，足以使多达4名技术员在里面工作。（"This interior environment"是介词"inside"的逻辑宾语）

**This river** is too dangerous <u>to swim</u> <u>in</u>.

这条河太危险了，（以至于）不能在里面游泳。（"This river"是介词"in"的逻辑宾语）

**The Atlantic** was too dangerous <u>for the Americans to send enough men</u> <u>and war materials</u> <u>across</u>.

当时大西洋太危险了，以至于美国人无法把足够的人员和战争物资运送过去。（"The Atlantic"是介词"across"的介词宾语。）

The author wishes to thank **the publishers** <u>for their assistance throughout</u> <u>and for being so easy to work</u> <u>with</u>.

作者要感谢各出版社在本书整个出版过程中所给予的帮助以及极为平易的合作。（"the publishers"是介词"with"的逻辑宾语。）

【改正后的句子】　**This material is very difficult to machine.**

### 例 208

【汉语原句】　为了发射无线电波，必须产生高频振荡。

【英语错句】　Sending out radio waves, high-frequency oscillations must be generated.

【错误分析】　(1)非谓语形式中一般只有动词不定式才能表示目的状语，所以这里应该使用"To send out radio waves"。(2)这个不定式的逻辑主语应该是"人"而不是"高频振荡"，这就成为"悬垂不定式"，当然在这里不会给读者产生歧义，在科技文中也经常见到的，但最好写成"it is necessary to generate high-frequency oscillations"或"we［one］must generate high-frequency oscillations"。

【改正后的句子】　**To send out radio waves, it is necessary to generate high-frequency oscillations.**


**例 209**

【汉语原句】　重要的是如何确定发射天线的方向图。

【英语错句】　It is important how to determine the transmitting antenna pattern.

【错误分析】　名词性不定式做主语时一般不用形式主语"it"这个句型，修改的方法有三种：一是把名词性不定式直接放在主语的位置；二是把名词性不定式改成由 how 引导的主语从句，这时就可以用形式主语 it 了，即改写成"how the transmitting antenna pattern is determined"；三是把名词性不定式前面的东西改成"what is important is"，这样语气就加强了，名词性不定式就成为句子的表语了。

【改正后的句子】　**What is important is how to determine the transmitting antenna pattern.**


**例 210**

【汉语原句】　为了减小振子间隔，就必须增加阵列的振子数。

【英语错句】　For reducing the element interval, the element number of the array must increase.

【错误分析】　(1)"for"这个介词处于句首时一般表示"对于，在……情况下"的含义，若要表示目的，英美人通常用不定式(用 for 短语的情况少见，当然也可以使用"for the purpose of …")。(2)"振子间隔"应该表示为"the interval between elements"。(3)"the element number"的含义往往是"振子号码"，而要表示"振子数"的话，应该使用"the number of elements"，这样就不会引起读者的误解。又譬如"the student number(学号)"而"the number of students(学生数)"。(4)"必须增加"的含义是"人们必须增加……"，所以采用了主动语态是错误的，应该采用被动句，即写成"the number of the elements of the array must be increased"，虽然句首不定式的逻辑主语并不是句子的主语"the number of the elements of the array"，但这种句子结构还是被英美科技人员广泛采用的，当然最好使用"it is necessary to increase the number of the elements of the array"。

【改正后的句子】　**To reduce the interval between elements, it is necessary to increase the number of the elements of the array.**

**例 211**

【汉语原句】　不仅温度和光影响导电率，而且给半导体加杂质也会使导电率变化很大。

【英语错句】　Not only temperature and light affect the conductivity, but the addition of impurities to semiconductors will also make it to change greatly.

【错误分析】　(1) 由于"not only"处于句首，所以句子应该发生部分倒装，因为句子的谓语动词为一般现在时"affect"，所以在"temperature"前应该加上助动词"do"。(2) 动词"make"需要的宾语补足语的不定式的标志"to"必须省去。

【改正后的句子】　**Not only do temperature and light affect the conductivity, but also the addition of impurities to semiconductors will make it change greatly.**

**例 212**

【汉语原句】　本文应用场匹配方法分析和计算了这种结构的色散特性和耦合阻抗，结果与 Karp 的实测值基本一致。

【英语错句】　In this paper, a field-matching approach has been used, and have obtained dispersion characteristic and coupled impedance of this structure. Theory gives good agreement with experimental results obtained by Karp.

【错误分析】　(1) 英文句中"have obtained"的主语不应该是"方法"，这里应该把逗号及"and have obtained"去掉，改写成"to analyze and calculate"，同时在其后面的名词前应加上定冠词。(2) "Theory gives"应改写成"The results obtained are in"。(3) "experimental results"改写成"experimental ones"好一些。由于这个句子是某论文的文摘中的一句话，所以一开头的"In this paper"是多余的，一般应省去。只有当文摘开头是讲某技术领域的状况等，而其论文主要就某个方面加以论述时才要用"in this paper"。例如："In recent years a substantial body of work on median filters and their generalizations based on order statistics has appeared in the literature of signal and image processing. This has provided impetus for the investigation of schemes for the implementation of such nonlinear filters. **In this paper** we present an algorithm for order statistic determination which operates in a number of steps less than or equal to the number of bits in the binary representation of the input, …"

【改正后的句子】　**A field-matching approach has been used to analyze and** calculate the dispersion characteristic and coupled impedance of this structure. The results obtained are in good agreement with the experimental ones given by Karp.

### 例 213

【汉语原句】 这些特点使电子对抗系统难于对这种信号截获、分析和干扰。

【英语错句】 These features make electronic counter-measure（ECM）systems difficult to intercept and analyze this kind of signal and jam.

【错误分析】 要表达本汉语句的含义，应该采用英语中常用的一个句型"…make it difficult for sb. to do sth."。例如："这些条件使得我们很难解这类方程。"应译成："These conditions make it difficult for us to solve this class of equation."

【改正后的句子】 These features make it difficult for electronic counter-measure（ECM）systems to intercept，analyze and jam this kind of signal.

### 例 214

【汉语原句】 本文介绍了 8098 微控制器与 386PC 机串行通讯的方法。

【英语错句】 This paper presents a method of 8098 microcontroller series-communicating with 386 personal computer.

【错误分析】 （1）在两个可数名词单数前均应该有一个不定冠词。（2）在"a method"之后应采用不定式复合结构的形式，即写成"for an 8098 microcontroller to series-communicate with a 386 personal computer"。当然在此也可以采用定语从句的写法，即："… a method by which an 8098 microcontroller series-communicates with a 386 personal computer."。不过这样句子并不简洁。

【改正后的句子】 This paper presents a method for an 8098 microcontroller to series-communicate with a 386 personal computer.

### 例 215

【汉语原句】 模拟结果表明，所有这两种方案是容易实现的。

【英语错句】 The simulation results prove that all these two schemes are easy to be implemented.

【错误分析】 （1）开头的"The"应该去掉。（2）"表明"一般用"show"

"indicate""illustrate"来表示，而"prove"则表示"证明"。（3）"所有这两种"不能用"all these two"表示，因为"all"这个形容词只能用于三者或三者以上的事物，这里应该使用"both the schemes"。（4）"to be implemented"应该改成主动形式"to implement"，因为这里应该使用"反射式不定式结构"。

【改正后的句子】　**Simulation results show that both the schemes are easy to implement.**

### 例 216

【汉语原句】　我们给出了三种算法，这些算法能明显地减少提起画笔的时间，从而提高绘画的效率。

【英语错句】　We give three algorithms. These algorithms can remarkably reduce the time for raising the pen, therefore improve drawing efficiency.

【错误分析】　（1）在"therefore"之前应加连接词"and"，不过这样修改后整个句子显得很松散，所以第二句可以改写成一个非限制性定语从句。（2）在"time"之后一般用动词不定式短语作后置定语。（3）句型改变后，把"therefore"改成"thus"或"thereby"，这样可用一个分词短语来表示结果状语。（4）"效率"一般与"raise"搭配使用。

【改正后的句子】　**Three algorithms are presented, which can remarkably reduce the time to raise the pen, thus raising the drawing efficiency.**

### 例 217

【汉语原句】　石英挠性加速度计抗恶劣环境能力的分析

【英语错句】　Analysis on the ability resisting bad environment for a quartz flexibility accelerometer

【错误分析】　这是一篇论文的标题，其中有两个错误：（1）在"analysis"之后一般跟"of"来引出逻辑宾语（即"分析"的对象），论文标题开头可以省去冠词。（2）在"ability"后应使用一种特殊的（即由"of"引出的）不定式复合结构来作后置定语，也就是"the ability of A to do B"这一模式。这里"A"就是"a quartz flexibility accelerometer"，而"do"就是"resist"。

【改正后的句子】　**Analysis of the ability of a quartz flexibility accelerometer to resist bad environment**

### 例 218

【汉语原句】　在运行这种 Pascal 程序时，用户只需输入应用问题的图形，就能在计算机上直接给出问题的解答。

【英语错句】 In running this PASCAL program, the user is only required to input the graph model of his applied problem, the solution is then given on the computer screen.

【错误分析】 (1) 表示"只需(做)……"时,英美人一般喜欢使用加强语气的句型 "All one needs to do is …",也有人用"It is only necessary to do …"。所以这里"用户只需……"可以写成"All the user needs to do is…"。(2) 由于涉及计算机屏幕,所以句子中的"给出"最好使用"displayed"而不用"given"。(3) 两句之间应加一个并列连词"and",或后一个句子缩成一个不定式短语,使整个句子显得比较简洁。

【改正后的句子】 **In running this PASCAL program, all the user needs to do is(to)input the graph model of his or her applied problem so as to get its solution displayed on the computer screen.**(注意:在句中的"his"之后加了"or her",主要是因为"the user"可能是男士,也可能是女士,这种表达方法在美国人讲话、写的文章或书籍中常见,这是对妇女与男士平等的一种表示,希望读者能注意这一点。)

**例 219**

【汉语原句】 该电路容易级联起来获得高阶电流模式滤波器。

【英语错句】 This circuit is easy to be cascaded to obtain high-order current-mode filters.

【错误分析】 本句有一个错误,主要受汉语影响。不少人见到汉语有"容易"就用"is〔are〕easy to(be done)"来表达,这是错误的,"is easy to be cascaded"应改写成"can be easily cascaded",这样句子显得简洁。(这里当然也可使用反射式不定式,即:"This circuit is easy to cascade so as to obtain …"。)

【改正后的句子】 **This circuit can be easily cascaded to obtain high-order current-mode filters.**

**例 220**

【汉语原句】 用这种算法获得结果只需要五次迭代,这比用传统的算法获得相同的结果的 250 次迭代少得多。

【英语错句】 It takes only 5 iterations for this algorithm to get result, which is much less than 250 iterations for the traditional algorithm to get the same result.

【错误分析】 （1）根据汉语原句及科技概念，"算法"本身并不能获得"结果"，只是人们利用算法来获得某种结果，因此不能用"算法"作为不定式"获得"的逻辑主语，应该写成"to obtain the result by using this algorithm"。（2）"which"在句中指的是"5 iterations"，所以连系动词在本句中应该用"are"。（3）"少得多"应该译成"far fewer"，因为这里是对可数名词复数进行比较，因此不能用"much less"。由英国学者 Michael Swan 所著的"Practical English Usage"说：对于可数名词来说，可以用"many more 或 far more"，但只能使用"far fewer"而一般不能说"many fewer"。（4）"用传统的算法获得相同的结果的"这一部分可以采用分词短语作后置定语写成"required to obtain the same result by using the traditional algorithm"或采用定语从句写成"it takes to obtain the same result by using the traditional algorithm"。

【改正后的句子】 **It takes only 5 iterations to obtain the result by using this algorithm，which are far fewer than 250 iterations required to obtain the same result by using the traditional algorithm.**

例 221

【汉语原句】 选择哪一种非线性函数值得进一步研究。

【英语错句】 It deserves to be further studied to select which nonlinear function.

【错误分析】 "to select which nonlinear function"是一种典型的中文式的英文表达法。修改的方法有两种，一是改成名词性不定式，即"which nonlinear function to select"，不过名词性不定式作主语时一般不使用形式主语"it"这一句型，应该把名词性不定式放在主语的位置；一是改成主语从句，写成"which nonlinear function should be selected"。如果把这样一句话"把哪一点选作为原点是没有关系的"写成英语的话，你会怎么写呢？有的人往往写成"It does not matter to select which point as the origin"。这犯的是与上述相同的错误，应该把它改写成"It does not matter which point is selected〔chosen〕as the origin"。

【改正后的句子】 **It deserves to be further studied which nonlinear function should be selected.**

例 222

【汉语原句】 为了能不受非线性失真的影响，Andreas Cyzlwink 提出了一种使用恒定包络导频符号的同步方法。

【英语错句】 In order to be robust to the nonlinear distortion，Andreas Cyzlwink proposed a synchronization using a constant envelop preamble.

【错误分析】 在句首作目的状语的动词不定式的逻辑主语通常应该是句子的主语，而本句句子的主语表示的是一个人，如果使用"to be robust to …"的话，等于是说"这个人为了不受非线性失真的影响"，这显然是不对的，因此应该把"to be robust to …"改成不定式复合结构"for the circuit to be robust to …"，意为"为了使电路不受…"。

【改正后的句子】 **In order for the circuit to be robust to the nonlinear distortion，Andreas Cyzlwink proposed a synchronization method using a constant envelop preamble.**

例 223

【汉语原句】 第一个衰减抽头与最后一个衰减抽头之比设定为 24 dB。

【英语错句】 The ratio of the first fading tap with the last fading tap is set to 24 dB.

【错误分析】 (1) "ratio"表示"A 与 B 之比"的搭配关系是"the ratio of A to B"(常见)或"the ratio between A and B"(不常见)，所以应该把句中的"with"改成"to"。(2) 在"is set to"之后应该加上"be"一词，"to be 24 dB"在此作主语补足语。

【改正后的句子】 **The ratio of the first fading tap to the last one is set to be 24 dB.**

例 224

【汉语原句】 我们必须把来自各传感器的信号按顺序发送给 eBOX。

【英语错句】 We are necessary to send the signals from the sensor to the eBOX in the order.

【错误分析】 (1) 为表示"我们必须"，一般有两个句型：一个是"we must (do …)"；另一个是"it is necessary for us to (do …)"而绝不能把它表示成"we are necessary to (do …)"。同样我们不能写成"we are easy[difficult] to (do …)"，而要写成"it is easy[difficult] for us to (do…)"。(2) "各传感器"应该是"each sensor"。(3) "按顺序"属于一个固定词组，即"in order"，在"order"前是不能加冠词的。

【改正后的句子】 **It is necessary for us to send the signals from each sensor to the eBOX in order.**

**例 225**

【汉语原句】　仅仅使用这种方法，还远远不足以模仿人类处理问题的能力。

【英语错句】　Only using this method is much deficient for simulating the ability with which the human beings deal with problems.

【错误分析】　(1) "远远不足"在此的确切含义应该为"far from being sufficient[enough]"。(2) 凡是与"sufficient[enough]"连用的动作应该用动词不定式来表示结果，即"to simulate …"。(3) 在"能力(ability)"后，应该使用"the ability of A to do B"结构，即"the ability of human beings to deal with problems"(注意在非特指的复数名词"human beings"前不加定冠词)。

【改正后的句子】　**Only using this method is far from being sufficient to simulate the ability of human beings to deal with problems.**

**例 226**

【汉语原句】　另一个问题是，如果在发射机中使用编码器，就要在接收机中加上译码器，因而这种方案在现有的系统中使用是不方便的。

【英语错句】　Another problem is that decoder will be added in receiver if coder is utilized in transmitter, thus this scheme is inconvenient to be used in existed systems.

【错误分析】　(1) 在"decoder""receiver""coder""transmitter"之前应该加上合适的冠词。(2) 由于人们是不愿意让这种情况出现的，所以最好使用虚拟语气，即"…would be added if … were utilized"。(3) 后面部分用"so that"引导一个结果状语从句要比"，and thus …"一个分句来得紧凑。(4) "to be used"应该改成"to use"，这属于"反射式的不定式结构"。(5) 由于"exist"为不及物动词，所以应该使用"existing"，当然也可使用"the systems available"。

【改正后的句子】　**Another problem is that a decoder would be added in the receiver if a coder were utilized in the transmitter so that this scheme is inconvenient to use in existing systems.**

**例 227**

【汉语原句】　有些颜色人们是难于给出精确名称的。

【英语错句】　Some colors are difficult to give accurate names for people.

【错误分析】　(1) 在科技文中，"people"最好换成"one"，这里"for one"应该放在"to give"之前，构成不定式复合结构作主语。(2) 在"names"后应该加介词"to"，本句也属于反射式不定式结构，这个"to"的逻辑介词宾语就是句子的主语"some colors"。

【改正后的句子】 **Some colors are difficult for one to give accurate names to.**

# Ⅱ、分 词

### 例 228

【汉语原句】　由这一行为产生的挑战能够用均匀设计法有效地克服。

【英语错句】　The challenges resulted from this action can effectively be overcomed by the uniform design method.

【错误分析】　(1)"effectively"要放在过去分词之前。(2) 由于"result"是一个不及物动词，所以要把"resulted"改成"resulting"。(3)"overcome"的过去分词是"overcome"。

【改正后的句子】 **The challenges resulting from this action can be effectively overcome by the uniform design method.**

### 例 229

【汉语原句】　为了解决由 $n$ 个传感器节点组成的无人值守传感器网路的数据存储可靠性问题，本文提出了一种新的分布式存储算法。

【英语错句】　To solve the data storage reliability problem of the unattended wireless sensor network consisted of $n$ sensor nodes，this paper proposed a new kind of distributed storage algorithm.

【错误分析】　(1) 由于"consist"是不及物动词，所以只能用"consisting"。(2)"propose"只能用一般现在时。(3) 要去掉"kind of"。

【改正后的句子】 **To solve the data storage reliability problem of the unattended wireless sensor network consisting of $n$ sensor nodes，this paper proposes a new distributed storage algorithm.**

### 例 230

【汉语原句】　适应性选择分解有助于加速算法收敛。

【英语错句】　Adaptive selecting decomposition is helpful to accelerate algorithm convergence.

【错误分析】 （1）"adaptive"应该用"adaptively"，因为它是修饰动名词"selecting"的。（2）这里的"to"是介词，所以其后面应该跟名词或动名词，类似的有"beneficial to …""conducive to …"等。

【改正后的句子】 **Adaptively selecting decomposition is helpful to accelerating algorithm convergence.**

**例 231**

【汉语原句】 让我们来看一下下一页上的那个例子。

【英语错句】 Let's now look at that example appeared on next page.

【错误分析】 （1）在科技写作中，一般是不用紧缩词的，所以应该把"Let's"写成"Let us"。同样"can't"应该写成"cannot"或"can not"；"we'll"应该写成"we shall"或"we will"；"it's"应写成"it is"或"it has"，等等。（2）"那个例子"在此不要写成"that example"，只要表示成"the example"就行了。（3）由于"appear"是一个不及物动词，所以不能用其过去分词作定语，而应该用现在分词"appearing"；同样我们不能用"existed""emerged""consisted of …""resulted from …""arised from …"等作定语，而要用它们的现在分词形式作定语。（4）在"next page"前应该加上定冠词，写成"the next page"；在"next"前只有在少数情况下不加定冠词，如："Thursday of **next week**"意为"下星期四"，"by this time **next year**"意为"到明年这个时候"。

【改正后的句子】 **Let us now look at the example（appearing）on the next page.**（这里把"appearing"放在括号里的意思是可以把它省略。）

**例 232**

【汉语原句】 该图片画出了一种种子，来自于一个人所持的袋子。

【英语错句】 The picture showed a kind of seeds which come from a bag holding in a person's hand.

【错误分析】 （1）"showed"应该改写成"shows"，也就是说，应该使用一般现在时。有人认为论文是在过去时间内写的，所以用了过去时，这想法是不对的，因为论文和书籍一样，只表示一种情况，因此均用一般现在时来表示，又如："This paper presents［introduces；discusses；describes；deals with …］"。（2）由于"hold"是一个及物动词，它与"bag"的关系属于被动的关系，所以应该使用它的过去分词"held"。

【改正后的句子】 **The picture shows a kind of seeds which come from a bag held in a person's hand.**

**例 233**

【汉语原句】　利用 PowerPoint 软件制作多媒体教学课件

【英语错句】　The Making Multimedia Teaching Software via PowerPoint

【错误分析】　（1）这是一篇论文的标题，修改的一种方法是把定冠词"the"去掉，用动名词短语作论文的标题。不过，在科技文章中，绝大多数的标题是用名词短语，只有极少数人用"on-短语"或动名词短语。但是要注意的是，科技论文的标题是不能用一个句子来表示的，而科普文章的标题可以用一个句子来表示，如"How a Computer Works"（计算机的工作原理），同时在标题中也不能出现各类从句。另一种修改方法是在"making"后加上"of"，这样"making"就成为"动名词名词化"了。（2）"课件"是"courseware"。（3）"利用 PowerPoint 软件"最好表示成"using the PowerPoint software"。

【改正后的句子】　**Making of Multimedia Teaching Courseware Using the PowerPoint Software**

**例 234**

【汉语原句】　有一些计算错误是由于对题目理解错误所引起的。

【英语错句】　There are a few miscalculations resulted from misunderstanding of problem.

【错误分析】　（1）"result from"（由……引起）这个词组属于不及物动词性质的词组，它是主动的形式、被动的含意，所以不能用它的过去分词形式作定语，而只能采用现在分词形式作定语；它当然也没有被动语态形式。（2）在"problem"前应该有冠词，因为它是可数名词单数，根据句意，在此要使用定冠词。

【改正后的句子】　**There are a few miscalculations resulting from （the） misunderstanding of the problem.**

**例 235**

【汉语原句】　控制扩散的因素有两个：（1）扩散信号。它可以是亮度信号或是运动匹配信号。（2）扩散信号的位置。它确定扩散过程的起始位置。

【英语错句】　There are two factors to control the diffusive process：(1) the diffusive signal. It can be brightness signal or motion-matching signal. (2) the position of the diffusive signal. It determines where the diffusive process starts.

【错误分析】　（1）根据英美人的习惯，在"factor"之后用现在分词做后置定语而很少用不定式。(2)本句是一个大句子,中间不该有句号存在,所以两个"It"前的句号均应改成逗号,并将"It"改成"which",这样就成了定语从句。(3)在(2)之前的句号应该改成分号。(4)在"brightness"和"motion"之前均应加一个不定冠词"a"。

**【改正后的句子】** There are two factors controlling the diffusive signal process：(1) the diffusive signal, which can be a brightness signal or a motion-matching signal；(2) the position of the diffusive signal, which determines where the diffusive process starts.

**例 236**

**【汉语原句】** 我们的方法与以前讨论同一问题的文章中所述的不同。

**【英语错句】** Our method is different from previous articles discussed the same problem.

**【错误分析】** (1)本句的比较对象不一致,主语是"方法",而"from"之后却是"文章"。(2)"discuss"为及物动词,其后面的"the same problem"为它的宾语,所以这时不能用过去分词而应该使用现在分词。(3)"以前的文章"用"previous articles"不好,因为"previous"一般在同一本书里或同一本论文集里表示"前面的"的意思时比较常用;"论文"显然在此指"科技[学术]论文",常用"paper"一词。

**【改正后的句子】** Our method is different from those presented (或 described) in the papers available (或 published before) on (或 discussing) the same problem.

**例 237**

**【汉语原句】** 已知常数 $K$,就能算出 $C$ 来。

**【英语错句】** Giving the constant $K$, $C$ can be calculated.

**【错误分析】** (1)表示"已知"的含义有两种方式,一种是用过去分词"given …",这表示这条件是被给定的,另一种是用现在分词"knowing …",表示"若知道了……的话"。(2)本句的主要部分应该使用主动形式,添加主语"one"或"we",因为前面作条件状语的分词短语的逻辑主语必须是句子的主语,如果不一致,则成为"垂悬分词(dangling participle)"或叫"无依着分词结构(unattached participial construction)"。这种结构在现代报刊英语和科技文体中还是屡见不鲜的,关键是不要引起意义上的含混。(但在参加各类考试时这一结构要算是错误的。)

**【改正后的句子】** **Given [Knowing] the constant $K$, one [we] can calculate $C$.** (或: ⋯, **it is possible to calculate $C$.**)

**例 238**

**【汉语原句】** 计算结果与其他数字方法相吻合。

**【英语错句】** The computing results agree with the other numerical methods.

**【错误分析】** (1)"计算结果"应该是"computed results"。(2)这里的比较对象不一致,所以在"with"后要写成"those (obtained) by ⋯"。

**【改正后的句子】** **The computed results agree with those(obtained) by the other numerical methods.**

**例 239**

**【汉语原句】** 计算机模拟结果验证了分析结论的正确性。

**【英语错句】** The computer simulation results validate the analyzed conclusions.

**【错误分析】** (1)句首的冠词可以去掉。(2)"the analyzed conclusions"应改成"the conclusions obtained through analysis"。

**【改正后的句子】** **Computer simulation results validate[verify] the conclusions obtained through analysis.**

**例 240**

**【汉语原句】** 现有的教科书没有提到这一点。

**【英语错句】** The now existed textbooks do not make mention of this point.

**【错误分析】** (1)"exist"属于不及物动词,所以是不能用过去分词来作定语的,在英语中只有少数几个不及物动词的表示状态的过去分词可作定语,如 "**fallen** leaves(落叶)""the **arrived** train(到达的火车)",因此这里应改成 "existing",不过"现有的教科书"一般也可译成"the textbooks available"。(2)凡用短语动词如 make use of、make mention of、pay attention to、take care of 等时,其否定式英美人一般使用"no"来否定词组中的名词。例如:

In this case **no use** has been made of that model.

在这种情况下并没有使用那个模型。

They have paid **no attention** to this phenomenon.

他们没有注意到这一现象。

【改正后的句子】　**The existing textbooks〔The textbooks available〕have made no mention of this point.**

### 例 241
【汉语原句】　他们对这一论题很感兴趣。

【英语错句】　They are very interesting in this topic.

【错误分析】　(1) 表示"人对……事物感兴趣"只能用过去分词"interested"(即 sb. is interested in sth.)，而不能用现在分词"interesting"，"东西对……人来说是感兴趣的"，则用"sth. is interesting to sb."。(2)"very"是不能修饰动词的，要改成"much"。

【改正后的句子】　**They are much interested in this topic.**

### 例 242
【汉语原句】　若把式(7)代入式(8)就得到 $a=b$。

【英语错句】　If Eq.(7) is substituted in Eq. (8), we obtain $a=b$.

【错误分析】　从语法上讲本句没有错误，但实际上英美科技人员在这种场合一般并不使用"if"从句来表示条件的，他们常用分词短语来表示这种条件，有时还可用动名词短语或名词短语作主语来表示条件的含意，其句型是：

① Substituting A in〔into〕B , we obtain〔have; get〕…

② Substituting A in〔into〕B gives〔yields; produces; results in; leads to〕…

③ Substitution of A in〔into〕B gives〔yields; produces; results in; leads to〕…

【改正后的句子】　**Substituting Eq.(7) in Eq.(8), we obtain $a=b$. (或：Substitution of Eq.(7) in Eq.(8) yields $a=b$.)**

### 例 243
【汉语原句】　导出的阻抗矩阵稀疏率高达 40%。

【英语错句】　Sparse ratio of the resulted impedance matrix is 40% high.

【错误分析】　(1) 句首应该加一个定冠词。(2)"导出的"应使用"resulting"，因为"result"作动词用时为不及物动词，或用形容词"resultant"，也可用"derived""developed"或"obtained"。(3) 表示"高达"这一程度应该使用

"as high as…"或"up to …"。

【改正后的句子】 **The sparse ratio of the resulting impedance matrix is as high as 40%.**

例 244

【汉语原句】 当 H 和 H′嵌在图 5 中时必须满足定理 4.1 和定理 4.6 的条件。

【英语错句】 H and H′ must satisfy the conditions of Theorem 4.1 and Theorem 4.6 when embedding them on Fig. 5.

【错误分析】 (1)"embedding them"应该写成"embedded",因为它的逻辑主语是句子的主语"H and H′",如果用现在分词短语的话则其逻辑主语是没有指出来的"人们"或"读者"了,这样从语法角度讲就错了。(2)很明显"on"应该改成"in"。

【改正后的句子】 **H and H′ must satisfy the conditions of Theorem 4.1 and Theorem 4.6（或：Theorems 4.1 and 4.6）when embedded in Fig. 5.**

例 245

【汉语原句】 本文介绍了用摩托罗拉公司制造的 MC145442 单片调制解调器多址通讯的一种新方法。

【英语错句】 This paper presents a new kind of approach to realize multi-point communication with MC145442 one-chip modem which is manufactured by MOTOROLA co.

【错误分析】 (1)去掉"kind of"。(2)"approach"后跟的"to"是介词而不是动词不定式的标志,这也是不少人在写作时经常搞错的,所以在"to"后面应该使用一个动名词短语。(3)在"with"后应该加上冠词"an"。(4)"which is"是多余的,可以直接用分词短语来作后置定语。(5)在"by"后要加定冠词。

【改正后的句子】 **This paper presents a new approach to realizing multi-point communication with an/the MC145442 one-chip modem manufactured by the MOTOROLA co.**

例 246

【汉语原句】 如果给定了返回函数是凸的,并且 Hessian 行列式是正的,那么该内点是一个最小值而不是一个最大值。

【英语错句】 The interior point is a minimum rather than a maximum, giving that the return functions are convex，and Hessian determinant is positive.

**【错误分析】** （1）表示给定的条件时实际上是由别人给的，所以应该使用"given"而不是"giving"。（2）由于本句的给定条件有两个，也就是说在"given"后有两个"保留宾语从句"，所以在"and"后应该加上从句引导词"that"。（3）在"人名＋名词"前一定要加上定冠词。

**【改正后的句子】** **The interior point is a minimum rather than a maximum，given that the return functions are convex，and that the Hessian determinant is positive.**

### 例 247

**【汉语原句】** 平均速度被定义为总的位移除以总的消逝的时间。

**【英语错句】** Average velocity is defined as the total displacement dividing the total elapsing time.

**【错误分析】** （1）"A 等于 B 除以 C"应该写成"A equals［is equal to］B **divided by** C."而"A 等于 B 乘以 C"应该写成"A equals［is equal to］B **multiplied by** C."。例如：

速度等于距离除以时间。

Speed is equal to distance **divided by** time.

电压等于电流乘以电阻。

Voltage equals current **multiplied by** resistance.

（2）"elapse(消逝)"这个词虽然是不及物动词，但习惯上要用其过去分词作定语，表示已完成的动作，所以在此应该用"elapsed"，又如"the hours worked（工时）"。

**【改正后的句子】** **Average velocity is defined as the total displacement divided by the total elapsed time.**

### 例 248

**【汉语原句】** 该模型能详细地描述发生故障的设备的性能。

**【英语错句】** The model can describe behavior of the malfunctioned equipment detailedly.

**【错误分析】** （1）在"behavior"前应该加定冠词。（2）"malfunctioned"应该改成"malfunctioning"，因为它是一个不及物动词。（3）"detailedly"应该改成"in detail"，因为并不存在"detailedly"一词。

【改正后的句子】　The model can describe the behavior of the malfunctioning equipment in detail.（更好的版本为：The model can give a detailed description of the behavior of the malfunctioning equipment.）

例 249

【汉语原句】　测量值进行实时显示，并每 30 秒钟更新一次。

【英语错句】　The measuring values are displayed real-timely and renewed for every 30 seconds.

【错误分析】　(1) 应该把"measuring"改成"measured"。(2) 这里的"实时"应该改为"in a real-time way"，因为没有"real-timely"一词。(3) 句中的"for"应该去掉，表示"每……"就用"every …"，其前面不能用任何介词。

【改正后的句子】　The measured values are displayed in a real-time way and renewed every 30 seconds.

## Ⅲ、动名词

例 250

【汉语原句】　本书所述的内容值得一读。

【英语错句】　The content of this book is worth to be read.

【错误分析】　(1) "The content of this book"的基本意思是可以的，但是如果要表达好一点的话，应该使用"what 从句"，即"What this book describes"或"What is described in this book"，英语中表示"论述"的词汇常见的有 describe、treat、present、cover、deal with、discuss 等。(2) 本句中的主要错误是"to be read"，而应该使用动名词的主动形式，即"reading"，这是"worth"这个词的特殊用法（如果使用"worthy"的话，就应该使用动词不定式的被动形式了，即"to be read"，比如："这个问题值得考虑。"应该写成"This problem is worthy to be considered."，当然也可写成"This problem is worth considering."，不过，实际上"worth"比"worthy"使用得广泛得多），整个句子就属于"反射式的动名词结构"，其定义与"反射式的动词不定式结构"类同，即"句子的主语是句尾动名词（或动名词短语末尾的介词）的逻辑宾语"这样的一种结构，例如：

**This computer** needs repairing.

这台计算机需要修理了。（"This computer"是"repairing"的逻辑宾语。）

**An old house** badly maintained would not be worth spending money on.

维护很差的旧房子就不值得在上面花钱了。("An old house"是介词"on"的逻辑宾语。)

**This work piece** is too complex to finish machining in such a short time.

这工件太复杂了，在这么短的时间内是加工不完的。("This work piece"是"machining"的逻辑宾语。)

【改正后的句子】 **What this book describes is worth reading.**

**例 251**

【汉语原句】 这相当于没有接收到功率。

【英语错句】 This corresponds of no power received.

【错误分析】 (1)表示"相应于"的词组是"correspond to"，所以应该把介词"of"改写成"to"。(2)在介词后的那部分表述有错，这里可以采用两种方式来改正它：一种是用一个名词"case"后跟一个定语从句，即"the case [situation] where [in which] no power is received"，不过这种句型似乎比较啰唆，第二种是使用"动名词复合结构"作介词宾语，即"no power being received"，这种方式在科技文中屡见不鲜。例如：

Automation is not a question of **machines replacing man.**

自动化并不是机器代替人的问题。

This is an example of **magnetism being converted into electricity.**

这是把磁转变成电的一个例子。

Tuned coupling minimizes the possibility of **undesired frequencies being radiated.**

调谐耦合使辐射不需要的频率的可能性，降到了最低的程度。

The solution of these independent equations often results in **certain of the currents being negative.**

解这些独立的方程式，往往会导致这些电流中有一些是负的。

A slow sampling rate is likely to lead to **the end station not receiving the bits correctly.**

抽样速度慢，有可能导致终端站不能正确地接收到这些比特。

The regular array of atoms in the lattice results in **there being certain sets of parallel and equally spaced planes in the crystal.**

原子在晶格中的规则排列，促使在晶体中形成了几组相互平行而等间隔的平面。

【改正后的句子】 **This corresponds to no power being received.**

例 252

【汉语原句】 这些数据经过处理后被立即送给数据库服务器。

【英语错句】 After processed, these data are sent to the database server immediately.

【错误分析】 本句的错误在"After processed"这一部分。"after"可以是介词，在它后面要跟名词或动名词，在此应该写成"After being processed"，由于"after"本身的词义，所以不必使用完成形式的动名词，这是科技文中常见的情况。"after"也可以是状语从句连接词，这时在它后面一般要用一个完整的句子，则应该写成"After they are processed"。有人可能会问："they"所代替的东西尚未出现，这时可以用代词吗？回答是肯定的，英语中的代词具有"代后"的功能（当然在译成汉译时，应把它所代的词先译出来，而把后面的名词译成代词或往往不需译出来）。例如：

Before it can work, **a computer** must be told what to do.

在计算机能够工作之前，必须被告知要干什么。（"it"代替了后面的"a computer"）

When they get hot, **all metals** melt.

所有的金属受热就会熔化。（"they"代替了后面的"all metals"）

Because of its simplicity, **the circuit** is widely used in power supplies.

由于该电路很简单，所以广泛地用在电源中。

In their study of electricity, **physicists** defined the electric field $E$ at a point in space as $E = F/q$.

物理学家们在研究电学时，把空间某一点处的电场强度 $E$ 定义为 $E = F/q$。

It cannot be proved here, but **the rational numbers do not take up all the positions on the line.**

虽然在这里无法证明，但有理数确实并不能占据该线上的所有位置。（"it"在此代替了"but"后的所述内容）

Surprising as it may seem, **this was the first direct verification of the reality of molecules.**

虽然似乎令人吃惊，但这毕竟是首次对于分子实际情况的直接证实。（"it"

在此代替了主句所述的内容）

【改正后的句子】 **After being processed, these data are sent to the database server immediately.**

例 253

【汉语原句】 在这种情况下，我们考虑把 $\xi_1$ 和 $\eta_3$ 连接起来。

【英语错句】 In this case, we consider to join $\xi_1$ and $\eta_3$.

【错误分析】 在"consider"之后如果跟一个动词，只能用动名词而不能用动词不定式。科技文中，常见的需要跟动名词的一些及物动词有 involve（涉及）、avoid（避免）、require（需要）、consider（考虑）、finish（完成）、suggest（启示）、facilitate（有助于）、resist（阻止）、imagine（想象）、imply（意味着）等。例如：

The final step involves **obtaining the inverse Laplace transform.**

最后一步涉及取拉氏反变换。

The word "ranging" means **measuring distance from a radar set to an object.**

"ranging"这个词的意思是测量从雷达机到某目标的距离。

Wire leads facilitate **inserting resistors into circuits.**

引线有助于将电阻器接入电路。

It is essential to avoid **making wrong connections of the polarities of this capacitor.**

必须避免接错这个电容器的极性。

These problems require **finding a maximum or minimum value of some function.**

这些题目需要求出某个函数的最大值或最小值。

Matter of any kind resists **being accelerated.**

任何物质均阻止被加速。

Now we may imagine **taking an air molecule at absolute zero and warming it up to room temperature a degree a time.**

现在我们可以设想取一个处于绝对零度的空气分子，然后一次一度地把它加热到室温状态。

Scattering implies **causing something to separate into different components.**

散射意味着使某东西分解成不同的分量。

Galileo's observations on the pendulum suggested **replacing these** dissipative mechanisms with resonant systems.

伽利略通过对钟摆的观察启示了可以用谐振系统来代替这些耗能机构。

【改正后的句子】 In this case, we consider joining $\xi_1$ and $\eta_3$.

# IV、动词及名词的数

例 254

【汉语原句】 所提出的非线性二阶偏微分方程能保持多得多的尖锐边缘。

【英语错句】 Proposed nonlinear second-order PDE leaves much more sharp edges.

【错误分析】 (1)主语前应该有冠词。(2)"leaves"应该是"maintains"。(3)"much"应该改成"many",因为这里的比较级是修饰一个可数名词复数"sharp edges"。

【改正后的句子】 The proposed nonlinear second-order PDE maintains many more sharp edges.

例 255

【汉语原句】 并讨论了这一方法与扫描法相比的优点。

【英语错句】 As compared with the scanning method, the advantages of this method has discussed also.

【错误分析】 (1)不少读者一看见"与……相比"就会想起使用固定短语"(as) compared with …",可是在"优点(advantage)"这个名词后地道的用法应该用其固定的搭配介词"over"来表示"与……相比"的含义。(2)本句主语是复数形式"advantages",所以谓语应该用"have"而不是"has"。(3)"讨论"在此只能用被动形式。

【改正后的句子】 The advantages of this method over the scanning method have also been discussed.

例 256

【汉语原句】 摩擦使得输入机器的功大于由机器所做的有用功。

【英语错句】 Friction make the work put into a machine be greater than the useful work accomplished by it.

**【错误分析】** （1）由于主语为"friction"，属单数第三人称，所以谓语要用"makes"，这是读者初学写作时经常犯的错误。（2）"be greater"中的"be"是多余的，因为形容词或名词或副词甚至介词短语均可作"make"要求的宾语补足语，所以不使用不带"to"的、由"be"为动词的不定式短语。例如：

We should make everything **clear**.

我们应该使所有问题都搞清楚。

One can make the iron bar **a magnet**.

人们能使铁棒成为一块磁铁。

This signal can make transistor Q3 **on**.

这个信号能使晶体管 Q3 导通。

They have made the piston **in the shape of a cylinder**.

他们把活塞做成了圆柱形。

**【改正后的句子】** **Friction makes the work put into a machine greater than the useful work accomplished by it.**

例 257

**【汉语原句】** 统计学习理论（SLT）表明在经验风险和学习机的能力之间必须要有恰当的平衡。

**【英语错句】** Statistical Learning Theory（SLT）show it necessary to have a right balance between the empirical risk and the capacity of learning machine.

**【错误分析】** （1）"show"应该用单数第三人称形式"shows"。（2）原作者对"show"的句型似乎不清楚，这个动词最常见的是跟有一个名词从句，另一个用法是用"as 短语"或"to be …"作补足语，它不能用于带有形式宾语"it"的句型，所以就本句而言，我们应该使用一个宾语从句，在"shows"之后应为"that it is necessary ＋动词不定式"。（3）这里的"有"实质上是指"存在有"的含意，所以不能用"have"这个动词，而应该使用"there be"句型，在此要使用它的不定式复合结构形式"for there to be"。下面再举几个例子。

It is possible **for there to be** a current in the circuit under such a condition.

在这种条件下电路中有可能<u>存在</u>电流。

In fact, it is not even necessary **for there to be** actual motion of either a wire or a source of magnetic field.

事实上，甚至并不一定要<u>存在</u>着导线或磁场源的实际运动。

The number of electrons or protons necessary in order **for there to be** a negative or positive charge equal to one coulomb is $6.2418 \times 10^{18}$.

为获得一库仑的负电荷或正电荷所需的电子数或质子数为 $6.2418 \times 10^{18}$ 个。（这是"in order to …"的不定式复合结构形式。）

（4）由于"theory"是可数名词单数，其前面应该有冠词，这里是特定的一个理论，所以应该用定冠词，同样，在"learning machine"前应该有冠词，这里该使用一个不定冠词"a"。（5）"恰当的"一般用"proper"表示。

**【改正后的句子】** The Statistical Learning Theory（SLT）shows that it is necessary for there to be a proper balance between（the）empirical risk and（the）capacity of a learning machine.

### 例 258

**【汉语原句】** 这些现象的存在及对它们的控制能力使得那些器件成为可能。

**【英语错句】** The existence and ability for controlling these phenomena makes those devices possible.

**【错误分析】** （1）在"existence"后应该加上介词"of"，它与动词"control"共用一个宾语"these phenomena"。（2）在名词"ability"之后应该使用动词不定式"to control"。（3）本句主语是两件明显不同的事，所以应该看成复数，因此"make"后不应加"s"。

**【改正后的句子】** The existence of and the ability to control these phenomena make those devices possible.

### 例 259

**【汉语原句】** 本文提出的模型及算法比传统的 BP 算法在性能上有一定的优越性。

**【英语错句】** The model and algorithm proposed in this paper has certain advantage of the conventional BP algorithm.

**【错误分析】** （1）由于句子主语涉及两个东西应该属于复数，所以谓语动词应该用相应的复数形式。（2）"一定的"在此可以用"to a certain degree [extent]"来表达。（3）"比……优越性"句中表达不正确，在此可以用形容词短语"superior to …"来表示。

**【改正后的句子】** The model and algorithm proposed in this paper are superior in performance to the conventional BP algorithm to a certain degree.

**例 260**

【汉语原句】　实验证实氧化层中的陷阱电荷对 FLOTOX 性能的退化起主要作用。

【英语错句】　The degration of FLOTOX is depended on the trap charges come from the tunnel oxide, which are proved by experimental results.

【错误分析】　本句写得很不好。（1）"退化"的名词是"degradation"。（2）在"FLOTOX"之前应该有冠词。（3）由于"depend on"是一个不及物动词性词组，所以不能有被动语态，应该改成"depends on"或"is determined by"。（4）"come"应该改成"coming"或"which come"。（5）逗号后由"which"引导的定语从句是修饰前面整个主句的，因此谓语应该是单数第三人称形式，即把"are"改成"is"。

【改正后的句子】　**The degradation of the FLOTOX depends on the trap charges coming from the tunnel oxide, which is proved by experimental results.** （更好的版本为：**Experiment shows that the trap charges in the tunnel oxide play a major （或：significant） role in the degradation of the FLOTOX.**）

**例 261**

【汉语原句】　然而在现实世界中，当有重大信息出现时，股票价格就会出现不连续的跳跃。

【英语错句】　However in real world, as significant information occur, the stock price occur discontinuous jump.

【错误分析】　（1）在"real world"前一定要有定冠词。（2）这里的"当"用"when"为好。（3）第一个"occur"应该为单数第三人称形式，且应该改为"appears"。（4）本句的主句部分是一个典型的中文式的英语句，实际上其主语应该是"discontinuous jumps"，而"股票价格"表示在哪一方面，应该写成"in stock price"。

【改正后的句子】　**However in the real world, when significant information appears, there will be discontinuous jumps in stock price.**

**例 262**

【汉语原句】　这种雷达通过多个频率与多个天线可实现各向同性照射。

【英语错句】　This kind of radar isotropically radiate by employing multiple frequencies and multiple antennas.

【错误分析】　（1）谓语没有使用单数形式，也就是动词词尾要加"s"。

（2）英美科技人员一般喜欢用名词来表示动作，也就是采用"radiation"，这时的谓语动词可采用"give off"。

【改正后的句子】　**This kind of radar can give off/realize isotropical radiation by employing（或：use of）multiple frequencies and multiple antennas.**

例 263

【汉语原句】　大家知道，当人们读中文字时，并不考虑其中的每一细节的。

【英语错句】　As it is known , when people read Chinese characters，they don't consider every detail in them.

【错误分析】　（1）在 as 引导的非限制性定语从句（它是说明整个主句的）中，"it"是多余的，应该去掉，它在此不代表任何东西，把"it"去掉后，关系代词"as"在句中起主语的作用，"as"意为"正如……那样"，这是不少读者常犯的错误。（2）在科技论文中，一般不出现"people"一词，往往用"one"来表示人们、我们、有人之意。例如：

If **one** wishes to send out radio waves，it is necessary to generate high-frequency oscillations.

如果人们（或：我们）想要发射无线电波，就必须产生高频振荡。（在本句中"one"作条件状语从句的主语。）

A bridge several times stronger than needed to carry its heaviest possible load serves no **one** better.

比需要来承受其可能的最大负荷结实好几倍的桥对我们是没什么好处的。（在本句中"one"作宾语。）

（3）从科技含义上来说，似乎"中文字"应该用单数形式更为合理，当"when"从句中用了"one"时，根据美国人的习惯，在主句中的人称代词应该用"he"或"she"（英国人可以只用"he"来表示）。（4）前面已提到过，在科技写作中英美人一般不使用缩略形式，如"don't"要写成"do not"，"it's"要写成"it is"或"it has"，"we'll"要写成"we shall［will］"，"can't"要写成"cannot"（英国式）或"can not"（美国式）等。

【改正后的句子】　**As is known，when one reads a Chinese character，he or she does not consider every detail in it.**

例 264

【汉语原句】　这个设备有它独特的一些优点。

【英语原句】　This equipment has a few advantage of all it's own.

**【错误分析】** （1）名词后表示复数的"s"往往漏写，这里应为"advantages"。（2）"it's"应改成"its"，不少读者经常习惯把"它的"写成"it's"，要知道 it's＝it is 或 it has。（3）"of"是多余的，"all its own(它特有的)"是作后置定语的。

**【改正后的句子】** **This equipment has a few advantages all its own.**

### 例 265

**【汉语原句】** 我和我的同事们想对 W.史密斯教授给予我们的大力帮助表示感谢。

**【英语错句】** I and my colleagues would like to express our thank to professor W. Smith for his great help.

**【错误分析】** （1）英美人表示人称代词时的词序为"第二人称→第三人称→第一人称"以表示客气，也就是说"我""我们"应写在后面，所以句中应写成"my colleagues and I"。（2）"感谢"若用名词"thank"来表示的话，应用复数形式，即"thanks"。（3）当教授、校长、总统、医生、同志等作为称呼的"头衔"时（即放在某人姓名之前时）其第一个字母一定要大写，其前面不得加冠词，即"Professor W. Smith"。

**【改正后的句子】** **My colleagues and I would like to express our thanks to Professor W. Smith for his great help.**

### 例 266

**【汉语原句】** 与目前国外资料中给出的天线必须平行地轴的理论相比，本文的方法可以实现更高的跟踪精度.

**【英语错句】** Comparing with the theory that the polar axis should be in parallel with the earth's axis, which was in the foreign literatures published in recent years, the method expressed in this paper will lead to higher accuracy.

**【错误分析】** （1）由于"Comparing"一词的逻辑主语应该是句子的主语"this method"，所以应该处于被动的关系，因此要用"Compared"。（2）"目前国外资料中给出的"这一概念若用定语从句表示的话谓语一般要用一般现在时"appears"而不应用过去时，也可以用分词短语来表示，放在"theory"之后。（3）名词"literature"属于不可数名词，在其后不得加"s"。（4）"expressed"应改成"given [presented; described; discussed…]"。

**【改正后的句子】** **Compared with the theory presented in the literature published abroad in recent years that the polar axis should be in parallel with the earth's axis, the**

**method given in this paper will lead to a higher tracking accuracy.**

**例 267**

【汉语原句】 标量检测会失去部分相位信息。

【英语错句】 Scalar detection loses some phase informations.

【错误分析】 (1)这里汉语的"会失去"并不是自己失去,而是"会导致失去",所以应该使用"will result in [will lead to]the loss of "来表达。(2)信息"information"这个词属于不可数名词,所以不得使用复数形式。

【改正后的句子】 **Scalar detection will result in the loss of some phase information.**

**例 268**

【汉语原句】 往往需要具有可变的分集增益来为不同衰减环境下的无线通讯提供恒定的服务质量。

【英语错句】 There often need to have variable diversity gain to provide constant Quality of Service(QOS)for wireless traffic in different fading environment.

【错误分析】 (1)"往往需要"可有两种表示法:① 使用"there is often a need for(doing …)"句型,② 使用"it is often necessary to(do …)"句型。(2)由于"gain"是可数名词,在其修饰语之前应该有冠词(在此应该使用不定冠词)。(3)由于"different"的关系,"environment"应该用复数形式。

【改正后的句子】 **It is often necessary to have a variable diversity gain to provide the constant Quality of Service(QOS)for wireless traffic in different fading environments.**

**例 269**

【汉语原句】 该新函数具有高阶导数,便于进行各种数学处理。

【英语错句】 This new function has high-order derivative which make it convenient for various kinds of mathematic disposal.

【错误分析】 (1)"高阶导数"应该用复数形式。(2)"which"引导的定语从句改为分词短语作状语为好,写成"…,(thus)making it convenient to(do …)"。(3)应该去掉"for",让"some kinds of …"作为"making"的宾语,而使"convenient"作宾语补足语。(4)这里的"处理"是不能用"disposal"一词的,因为它的意思是"清除;去掉",而应该使用"treatment"或"manipulation"。

（5）"mathematic"一般应该改用"mathematical"。

**【改正后的句子】** **This new function has high-order derivatives，（thus）making it convenient to perform various kinds of mathematical treatment.**

例 270

**【汉语原句】** 这种新算法的特点是简单而客观。

**【英语错句】** The feature of this new algorithm is simple and objective.

**【错误分析】** （1）表示"特点"的同位语不能用形容词而只能用名词，应写成"simplicity and objectivity"，这是科技人员写作时常犯的一种错误，好像语法上很正确，而实际上，逻辑是错误的。（2）由于有两个特点，所以主语应该用复数形式，连系动词也相应改成"are"。

**【改正后的句子】** **The features of this new algorithm are simplicity and objectivity.**

例 271

**【汉语原句】** 方程(10)和(17)具有负的根是可能的，我们认为这等同于信号的相位有 180°的偏移。

**【英语错句】** It is possible that Eq.（10）and（17）have negative roots，we think that it is equivalent to the situation that the phase of the signal has 180° shift.

**【错误分析】** （1）"Eq."应该用复数形式"Eqs."，这是不少读者不清楚的。（2）为了使句子紧凑，把逗号前后两部分通过一个非限制性定语从句联系起来，而把"we think"作为插入句来处理，即"which we think is equivalent to the situation"。（3）在"situation"后应该使用由"where（＝in which）"引导的定语从句，而不少读者往往错用"that"引导同位语从句。（4）由于"shift"在此是非特指的可数名词单数，所以在其前面应该加有不定冠词。

**【改正后的句子】** **It is possible that Eqs.（10）and（17）have negative roots，which we think is equivalent to the situation where the phase of the signal has a 180° shift.**

例 272

**【汉语原句】** 性接触率被定义为一个 HIV 携带者每年与他人可能有的性接触的次数。

**【英语错句】** The sexual act rate is defined as the times of sexual act a

HIV carrier may have with the other people per year.

【错误分析】　（1）对一个术语进行定义时，该术语前不用冠词。（2）"次数"应该表示为"the number of times"，而"times"仅意为"次"或"倍"，而"数"这一概念并没有表示出来，这是不少读者经常出错的地方。（3）后面的"sexual act"应该用复数形式。（4）在"HIV"之前应该用"an"。（5）由于"其他人"并没有特定，所以在其前面不该用定冠词。（6）由于定语从句是修饰"the number of times"，所以其引导词可以省去，也可用万能关系副词"that"或用"by which"。

【改正后的句子】　Sexual act rate is defined as the number of times（that/by which）an HIV carrier may have sexual acts with other people per year.

# Ⅴ、定语从句（包括"as"从句）

**例 273**

【汉语原句】　本文的发现以下面的简要讨论来展示。

【英语错句】　The findings of this paper are illustrated with the brief discussion as follows.

【错误分析】　（1）"of"改为"in"比较好。（2）"as follows"意为"如下"，一般作状语，所以应该把"as"改为"that"。

【改正后的句子】　The findings in this paper are illustrated with the brief discussion that follows.

**例 274**

【汉语原句】　这些方程是 $x$ 和 $t$ 均为零的特殊情况。

【英语错句】　These equations are special cases that $x$ and $t$ both are zero.

【错误分析】　（1）修饰"case"的定语从句应该由"where"或"in which"来引导。（2）作主语的同位语时"both"应该放在 is［are］之后，"all"的用法也是如此，如："他们都是大学生。"要译成"They are **all** college［university］ students."。

【改正后的句子】　These equations are special cases where $x$ and $t$ are both zero.

**例 275**

【汉语原句】　这些是中国曾制造过的最大的飞机。

**【英语错句】**　These are the largest aircrafts which were ever manufactured in China.

**【错误分析】**　（1）"aircraft"一词是单复数同形，这一点要特别注意。（2）由于先行词受到形容词最高级的修饰，所以先行词后面的定语从句根据传统语法规则应该用"that"来引导而不能用"which"，虽然它们均可用来引导限制性定语从句修饰表示事物的名词，但在以下几种情况下，就要用"that"：

①　先行词为"all，everything，nothing，something，anything，little"等不定代词时，例如：

Anything **that is hot** radiates heat.

任何热的东西均辐射热量。

All **that one should do** is push the button.

我们只需要按一下这个按钮。

②　先行词被序数词修饰时，例如：

The first thing **that will be done** is to measure the voltage across this component.

要做的第一件事是测出该元件两端的电压。

③　先行词被形容词"only，no，very，any"等修饰时，例如：

The only measure **that one can take in this case** is to connect a capacitor across $R_1$.

在这种情况下我们所能采取的唯一措施是在 $R_1$ 两端并接一个电容器。

④　先行词被形容词最高级修饰时，例如：

Computers are the most efficient assistants **that man has ever had**.

计算机是人类曾有过的效率最高的助手。

In practice, the best **that the analog computer** can achieve is an accuracy of about 1 part in 10,000.

实际上，模拟计算机所能达到的最高精度大约为万分之一。

（3）在定语从句中由于用了副词"ever"，是指到目前为止的情况，因此谓语应该使用现在完成时，即"have been manufactured"。

**【改正后的句子】**　These are the largest aircraft **that have ever been manufactured in China**.

**例 276**

**【汉语原句】**　下面这个概念是神经系统所基于的基本原理之一。

**【英语错句】**　The concept as follows is one of basic principles based by the neural system.

【错误分析】 （1）"as follows"是不能作定语从句的，要把"as"改成"which"。（2）主要错误在于"based by the neural system"，通常应该采用一个定语从句，即写成"on which the neural system is based"。（3）在"basic"之前应加一个定冠词"the"。（4）由于这里是指一般的"神经系统"，所以要用复数形式"neural systems"。

【改正后的句子】 **The concept which follows/The following concept is one of the basic principles on which neural systems are based.**

例 277

【汉语原句】 图 3 画出了一根传输线，其长度为 $l$，而其特性阻抗为 50 Ω。

【英语错句】 Fig.3 shows a transmission line which length is $l$ and the characteristic impedance is 50 Ω.

【错误分析】 （1）本句主要是定语从句写错了，这里"which"应改写成"whose"或"the length of which"。（2）"特性阻抗为 50 Ω"也应该是属于定语从句，所以同样应该把"the"改成"whose"或"the characteristic impedance of which"，更简洁的写法是用"with 短语"，即"with a length of $l$ and a characteristic impedance of 50 Ω"。

【改正后的句子】 **Fig. 3 shows a transmission line with a length of $l$ and a characteristic impedance of 50 Ω.**

例 278

【汉语原句】 空气中的导体可以被提升到的最大电位，受到了空气分子变成电离因而空气变成了导体的限制。

【英语错句】 The maximum potential for which a conductor in air can be raised is limited by that air molecules become ionized, and hence the air becomes a conductor.

【错误分析】 （1）"for which"应改成"to which"，因为"to"是从句中的动词"raise"所要求的。（2）前面讲过，"that"是不能引导介词宾语从句的，应该在"by"后面加上"the fact"。（3）把"and hence"改成"with the result that"使句子显得更好些。

【改正后的句子】 **The maximum potential to which a conductor in air can be raised is limited by the fact that air molecules become ionized with the result that the air becomes a conductor.**

**例 279**

【汉语原句】　最后，本文给出了有待于解决的几个问题。

【英语错句】　Finally, this paper presents several problems remain to be solved .

【错误分析】　本句的错误在于一个句子中出现了两个谓语动词"presents"和"remain"，后者应该属于一个定语从句的，所以可以在"remain"之前加一个关系代词"that"或"which"，这一句型在论文的文摘中经常使用。（当然这里也可使用"remaining to be solved"，不过这种用法并不常见。）

【改正后的句子】　**Finally, this paper presents several problems which remain to be solved.**

**例 280**

【汉语原句】　本文提出了点到点（P2P）模式下协同工作安全群组通信模型，它允许用户动态加入和退出系统。

【英语错句】　Group secure communication model of CSCW in P2P network is proposed. This model permits the users to join and quit the system dynamically.

【错误分析】　（1）本句开头应该加一个不定冠词。（2）在"P2P network"之前应该加定冠词。（3）把". This model"改成", which"，这样使句子结构更紧凑。（4）在"users"前不要加定冠词，因为这里的"用户"并没有特指。

【改正后的句子】　**A group secure communication model of CSCW in the P2P network is proposed, which permits users to join and quit the system dynamically.**

# Ⅵ、"what"从句

**例 281**

【汉语原句】　本文所讲的内容对通讯工程师来说是很感兴趣的。

【英语错句】　The content that this paper talks about is of great interest to communications engineers.

【错误分析】　（1）"talks about"是典型的中文式的英文；英美人常用 describe、treat、discuss、deal with、cover、present 等来表示这里的所谓"讲"的含义。（2）虽然"content"表示了"内容"这一含义，但英美人常用"what 从句"来表示,应该改写成"what this paper describes〔covers; discusses; deals with; …〕",当然也可以用被动形式,即"what is described〔covered; treated; discussed; dealt with; …〕 in this paper","what"从句在英语中使用得非常广泛,它可以表达许多不同的含义,例如(注意"what"的不同译法):

**What I have said above** is not necessarily correct.

我上面所讲的不一定正确。

**What is described in this chapter** is of great importance.

这一章所讲的内容非常重要。

**What we need** is an oscilloscope.

我们所需要的是一台示波器。

Its actual direction is opposite to **what has been assumed**.

其实际方向与我们假设的方向相反。

Energy is **what brings changes to materials**.

能量是使物质发生变化的东西。

This is close to **what has been observed**.

这接近于人们观察到的情况。

Gas takes the shape of **what is holding it**.

气体呈现存放它的容器的形状。

The turning of the earth on its own axis is **what makes the change from day to night**.

地球的自转,是引起日夜变化的原因。

【改正后的句子】 **What this paper describes is of great interest to communications engineers.**

## 例 282

【汉语原句】 尚不清楚到底在什么条件下能够使用这种方法。

【英语错句】 It is not clear yet that it is under what conditions that this method can be used.

【错误分析】 (1)"yet"之后的"that"应该去掉,因为后面由连接代词"what"引导主语从句了,这也是常见的错误。(2)虽然句中的强调句型"it is that"是用来强调在主语从句中作状语的介词短语"under what conditions"的,但是由于"what"引导主语从句,所以要把它放在从句之首,变成了"under what conditions it is that …"。例如:

No one knows what **it is that** composes the field itself.

没有人知道到底是什么东西构成了场本身。("what"引导宾语从句。)

It is still not clear why **it is that** this is so.

仍然不清楚到底为什么会是这样的。("why"引导主语从句。)

Fig. 1-3 shows how **it is that** a coil of wire induces a current in another coil when the current in it is turned on or off.

图 1-3 画出了当一个线圈中的电流通断时到底是如何在另一个线圈中感应出电流来的(情况)。("how"引导宾语从句。)

By convention，we call whatever **it is that** the rubber rod possesses by virtue of having been stroked with the fur negative electrical charge.

习惯上，我们把那种由于用毛皮摩擦而在橡胶棒上所带的电荷称为负电荷。("whatever"引导宾语从句。)

What **is it that** this sequence "tends to"?

这个数列到底"趋于"什么值？("what"引出一个特殊疑问句，所以"it is that"变成了"is it that"。)

About how many elements **is it that** make up most of the substances we meet in everyday life?

到底大约有多少元素构成了我们在日常生活中所见到的大多数物质呢？("how"引出一个特殊疑问句。)

**【改正后的句子】** It is not clear yet under what conditions it is that this method can be used.

**例 283**

**【汉语原句】** 用户与 IN 网络之间的交互作用可能不同于转式 POTS 电话机中的情况。

**【英语错句】** The interaction between user and IN network may be different from that happens in the rotary POTS telephone set.

**【错误分析】** 这里的主要错误在于使用了"that"而应该使用"what"。前面提到过，除个别固定词组外，"that"是不能引导介词宾语从句的，所以在介词后不能用"that"引出一个从句。

**【改正后的句子】** The interaction between user and IN network may be different from what happens in the rotary POTS telephone set.

**例 284**

**【汉语原句】** 公共照明所消耗的电能是巨大的，大约为整个照明系统的 32％。

【英语错句】 The consumed electricity of public lighting is tremendous, about 32％ out of the whole lighting system.

【错误分析】 （1）本句的主语部分"The consumed electricity of public lighting"写得不好，有"中文式的英语"之嫌，应该表示成"The amount of the electric energy consumed by public lighting"。（2）在"32％"之后那一部分写错了，应该用"what 从句"来表示，即："of what the whole lighting system consumes"（"of"经常可以省去的）。

【改正后的句子】 **The amount of the electric energy consumed by public lighting is tremendous, about 32％ (of) what the whole lighting system consumes.**

# Ⅶ、状语从句及比较句型

例 285

【汉语原句】 强加于伺服马达并使控制系统的性能严重恶化的那些因素已经被消除了，因而很有可能提高系统的性能。

【英语错句】 Those factors which impose on the servo motor seriously worsen the performance of the control system have been eliminated. Therefore, there is a great opportunity to improve the system performance.

【错误分析】 （1）由于"impose"是及物动词，所以在此应该用被动语态。（2）英语句的第二部分是第一部分产生的结果，因此最好采用一个结果状语从句来表示，这样整个句子显得比较紧凑。（3）"很有可能……"一般常用"it is quite possible to（do）…"来表示。

【改正后的句子】 **Those factors which are imposed on the servo motor and seriously worsen the performance of the control system have been eliminated so that it is quite possible to improve the system performance.**

例 286

【汉语原句】 铜的导电率比铁高。

【英语错句】 The conductivity of copper is higher than iron.

【错误分析】 本句由于受到了汉语句中省略现象的影响而出现了错误，即比较对象不一致，一个是"conductivity"而另一个则是"iron"，应该在"iron"之前加上"that of"才对，这种错句经常出现。

【改正后的句子】　The conductivity of copper is higher than that of iron.

例 287

【汉语原句】　在这个实验室中，这台仪器比其他的来得灵敏。

【英语错句】　In this laboratory, this instrument is more sensitive than any one.

【错误分析】　使用比较级时，主句的主语是绝不能包含在比较从句的比较对象之中的，这也是我国读者常犯的一个错误，因此本句中的"any one"是错误的，应改成"any other one"或"any one else"。

【改正后的句子】　In this laboratory, this instrument is more sensitive than any other one（或：any one else）.

例 288

【汉语原句】　一个外围结构的笔画数应尽可能的少。

【英语错句】　The number of strokes of a peripheral structure should be as less as possible.

【错误分析】　在汉语句中的"少"是修饰说明"数"的，在英语中说明"number"的形容词只能是"large""great"或"small"，所以不能用"little"，更不能用其比较级"less"（因为"as…as possible"只能用于形容词或副词的原级），这里应该用"small"。

【改正后的句子】　The number of strokes of a peripheral structure should be as small as possible.

例 289

【汉语原句】　声音传播的速度比光低得多。

【英语错句】　The sound travels less faster than light is.

【错误分析】　(1)"sound"在此表示一般性的含义，并没有特指，所以在其前面不用冠词。(2)表示比较的程度"得多"应在比较级前加上一个副词，如："much""far""well""greatly""considerably""appreciably""significantly"或"substantially"，也可以加上"a lot"或"a great deal"。(3)用了"less"后"fast"只能用原级而不得用比较级了。(4)"is"应改成代动词"does"，它代替动词"travels"，当然也可以省去这个代动词。

【改正后的句子】　Sound travels much less fast than light（does）.

　　例 290

【汉语原句】　物体越热，它辐射的能量就越多。

【英语错句】　The hoter the body，the more the energy it radiates.

【错误分析】　(1)"hot"这个词属重读闭音节的形容词且该音节最后只有一个辅音字母，在构成比较等级时要双写最后一个辅音字母，变成"hotter"（但"thick"在构成比较级时其最后一个字母就不能双写了，因为其后有两个辅音字母"ck"）。(2)逗号后的那个英语表达式虽从语法结构来说是正确的，它完全符合汉语的表达，但英美科技人员一般不采用这种说法，而常用"the more energy it radiates"。表面上看这两种方式好像只差一个定冠词"the"，实际上其句子结构是大不一样的，前者的表达式中主语是"the energy"，"it radiates"为省去了关系代词"that"或"which"（它在从句中作为及物动词"radiates"的宾语）的一个定语从句修饰"energy"，谓语动词为"will be"或"is"（在句中已省去了），"the more"是表语；而在后者中，主语为"it"，谓语为"radiates"，宾语为"the more energy"（由于形容词比较级"the more"修饰"energy"，所以把"energy"一词提到句首了）。这是常见的一种句型，希望读者要熟悉它。类似的句子有：

The harder you work，**the greater progress** you will make.

你工作越努力，你取得的进步就越大。

The larger the resistance，**the more energy** it consumes.

电阻越大，它消耗的能量就越多。

【改正后的句子】　**The hotter the body，the more energy it radiates.**

　　例 291

【汉语原句】　高维空间内的复杂模式识别题比低维空间更有可能线性可分。

【英语错句】　The complex pattern recognition problem in the high dimensional space is more possible to be linearly separable than in the lower dimensional space.

【错误分析】　(1)"possible"作表语时，其相应的主语一般只能是一件事，而不能是一个人或一样东西，所以这里应该把它换成"likely"，这个词可以后跟动词不定式。例如：

This phenomenon is **likely** to occur in such a condition.

在这种条件下很可能会出现这一现象。

He is not **likely** to come.

他不见得会来。

（2）本句的比较对象不一致，应该在"than"后加一个代词"that"为好。（3）为了前后对应，最好把"high"改成"higher"；或者把"lower"改为"low"。

【改正后的句子】 The complex pattern recognition problem in the higher dimensional space is more likely to be linearly separable than that in the lower dimensional space.（更好的版本为：The complex pattern recognition problem is more likely to be linearly separable in the higher dimensional space than in the lower dimensional space.）

**例 292**
【汉语原句】 在这种情况下，HSSVM 与 SVM 具有同样好的推广性能。
【英语错句】 In this case, HSSVMs have a generalization performance as well as SVMs.
【错误分析】 "as well as"在此用错了，应该写成"as good a generalization performance as SVMs(do)"，这里注意要出现不定冠词"a"的特殊位置；或者写成"as good as that of SVMs"，不过这种写法没有前面那种好。
【改正后的句子】 **In this case, HSSVMs have as good a generalization performance as SVMs（do）.**

**例 293**
【汉语原句】 滤波器的齿数变化对滤波器通带性能的影响比结构尺寸变化产生的影响大。
【英语错句】 The variation of the filter's tooth number more affects the performance of its passband compared with the variation of its dimensions.
【错误分析】 （1）本句属于比较级句型，所以不该使用"compared with …"这一短语，应把它改成"than"，并且把"more"直接放在"than"之前为好。（2）"齿数"应该用"the number of teeth"，因为"tooth number"有可能被误解成"齿号"。（3）把"影响"改成名词则更好，写成"… has a greater effect on … than …"。
【改正后的句子】 **The variation of the number of the filter's teeth has a greater effect on the performance of the filter's passband than the variation of its dimensions.**

**例 294**
【汉语原句】 异常程度越大，残留的失配就越大，曲线就上扬。

【英语错句】　Larger is abnormal extent, larger is the residual mismatch and the curve goes up.

【错误分析】　（1）本句前一部分属于"越……，越……"句型，其公式应该是" the＋比较级……，the＋比较级"，所以两个"larger"前应该加上"the"，这是一些读者写作时往往遗忘掉的；该句型的第一部分中的连系动词"is"是不倒装的，也就是说应该放在这一部分的主语"abnormal extent"之后（当然也可以把"is"省去），同时该主语前应该加上定冠词"the"。（2）为了使句子更紧凑，可以把"and the curve goes up"用一个"with 结构"来代替。

【改正后的句子】　**The larger the abnormal extent（is），the larger is the residual mismatch，with the curve going up.**

例 295

【汉语原句】　这种新算法的解释与 Gerchberg 算法是很不相同的，但新算法的计算结果与 Gerchberg 算法相同.

【英语错句】　The interpretation for the new algorithm is much different from the Gerchberg algorithm but the computing results of the new algorithm are same as what results from the Gerchberg algorithm.

【错误分析】　（1）在名词"interpretation"后应该用介词"of"来引出它的逻辑宾语（即"动宾关系"）。（2）"much"一般不能用来修饰形容词，而应该使用"very"或"quite"来表示"很"的含义。（3）"from"之后的比较对象与主语不一致，应该写成"that of the Gerchberg algorithm"。（4）"计算结果"应写成"computed［computational］results"，"模拟（仿真）结果"可译成"simulation results"，"实验结果"可译成"experimental results"，"分析结果"可译成"analytical results"等。（5）在"same"前要有定冠词。（6）在"results"后的"of"应该用"by"，因为实际上是"用新算法计算出来的结果"。（7）"what results from …"应改成"those（computed）by …"。

【改正后的句子】　**The interpretation of the new algorithm is quite different from that of the Gerchberg algorithm，but the results computed by the new algorithm are the same as those（computed）by the Gerchberg algorithm.**

例 296

【汉语原句】　这个信号所含的信息比那个多四倍。

【英语错句】　This signal includes four times more information than that one does.

【错误分析】 （1）"includes"应该改成"contains"。（2）如何表示"倍数的增减"，这是许多读者不清楚的一个问题，往往受到了汉语的影响。普通英语和科技英语的表达法是一致的，我们可以用公式表示如下：

**n times** ＋ 比较级 **than** … ＝ **n times as** 原级 **as** …

简单地说，"英译汉减一倍；汉译英加一倍"。因此本句中的"four"应改写成"five"。例如：

This book is **four times** thicker than that one（is）.

这本书比那本厚三倍。

XVI is used to express the number of 16. To express a number **a thousand times** larger，the Romans put a line above that number.

XVI 用来表示数"16"。为了要表示其一千倍那么大的（即：比它大 999 倍的）一个数，罗马人就在那个数的上方加一横杠。

The average $N_u$ is found to equal to 23.5 or nearly **4.5 times** greater than the ideal $N_u$ of 5.38.

人们发现平均 $N_u$ 值为 23.5，也就是比理想的 $N_u$ 值（5.38）几乎大了 3.5 倍（或：几乎是理想 $N_u$ 值的 4.5 倍）。

The wavelength of this musical note is 7.8 ft，over **three times** longer than the wavelength of the same note in air（2.5 ft）.

这个音符的波长为 7.8 英尺，比同一音符在空气中的波长（2.5 英尺）长两倍多。

【改正后的句子】 **This signal contains five times more information than that one（does）.**（或：**This signal contains five times as much information as that one（does）.** ）

# Ⅷ、语态和时态

例 297

【汉语原句】 只有通过对该系统性能的研究，我们才能了解它的优点。

【英语错句】 Only through studied the performance of the system，its advantages can be understood.

【错误分析】 （1）"through"是介词，所以后面不能用过去分词"studied"，英美科技人员习惯上常使用"through＋名词＋of＋…"或"by＋动名词［动作性名词＋of］＋…"来表示"通过……"，所以这里应改成 "through the study of

…"或"by studying …"。（2）本句由于以"only＋状语"开头，所以后面应发生部分倒装现象，也就是说"can"应该倒装在句子主语之前。（3）由于"研究"的主语是人，所以本句应该使用主动语态为好。

【改正后的句子】 **Only through the study of the performance of the system, can one（或：we）understand（或：appreciate）its advantages.**

### 例 298

【汉语原句】 这模块是由五部分构成的。

【英语错句】 This module is consisted of five parts.

【错误分析】 不少人经常把动词"consist"当成是一个及物动词，因而写出了错误的句子，应该改成"consists of"。若用被动句，则要使用另一个词组"be composed of"或"be made up of"。不过多数英美人常用"consist of"这一词组，因为它比较简练。

【改正后的句子】 **This module consists of five parts.**

### 例 299

【汉语原句】 该系统的测试是在香港的一家工厂进行的，结果令人满意。

【英语错句】 **The test of the system has taken at a factory in Hong Kong with results satisfied.**

【错误分析】 （1）"测试"自己不会进行的，只能采用被动语态句型。（2）"进行测试"不是"take a test"而是"conduct a test"。（3）"with results satisfied"似乎采用了一个"with 结构"的形式，但特别要注意的是过去分词与现在分词使用上的区别：若说明人得到了满足则要用过去分词 satisfied，若某东西使人满意，则要用现在分词 satisfying 或形容词 satisfactory。又如："这位教授对该论题很感兴趣"应写成"The professor is quite interested in that topic."（不少人往往在此写成了 interesting，这是绝对错误的），而如果说"该论题对这位教授来说是很有趣的"应写成"That topic is quite interesting to the professor."。（4）为了使句子简洁起见，"结果令人满意"这一含义可以用一个"with 短语"表示，写成 with satisfactory results 而不必采用一个"with 结构"。（5）本句的时态应该用一般过去时。

【改正后的句子】 **The test of the system was conducted at a factory in Hong Kong with satisfactory results.**

### 例 300

【汉语原句】 已表明：随着发送天线数的增加，该下限会愈加紧密。

【英语错句】　It has shown that this low bound becomes tighter with the number of transmitter antennas growing.

【错误分析】　(1)"已表明"应该写成"It is shown"或"It has been shown"。(2)"下限"应该是"lower bound"。(3)"发射天线"一般为"transmitting antenna"。(4)"……数的增加"应写成"the increasing number of …"或"an increase in the number of …"

【改正后的句子】　**It is shown that this lower bound becomes tighter with the increasing number of transmitting antennas.**

### 例 301

【汉语原句】　上面所讨论的表明,这种可能性将大一倍。

【英语错句】　What has discussed above shows that this possibility will be one time greater.

【错误分析】　(1)在主语从句中应该使用被动语态。(2)前面介绍过,表示倍数时,读者应该记住"英译汉减一倍;汉译英加一倍",所以这里"将大一倍"应该写成"will be two times[twice]greater"。

【改正后的句子】　**What has been discussed above shows that this possibility will be two times greater.**

### 例 302

【汉语原句】　我们发现,在这种条件下最大跨导提高 8.9%,肖特基栅极的反向电流减小两个数量级。

【英语错句】　We find that under this condition the maximum transconductance raises 8.9% and the reverse leakage of schottky gate reduces to two orders lower.

【错误分析】　(1)"raise"和"reduce"是及物动词,所以在这里应该用被动形式。(2)按照现在这个句子,在"the reverse leakage"前应该加一个宾语从句连接词"that"。(3)在"schottky gate"前应该加上定冠词。(4)"and the reverse leakage …"改用"with 结构"为好。(5)"to two orders lower"应该为"by two orders of magnitude"。

【改正后的句子】　**We find that under this condition the maximum transconductance is raised by 8.9%, with the reverse leakage of the schottky gate reduced by two orders of magnitude.**

**例 303**

【汉语原句】 这本书的序言写得很好。

【英语错句】 The preface of this book was well written.

【错误分析】 (1)"preface"后跟"to"而不用"of",这属于固定的词汇搭配关系。(2)本句主要叙述一种状态而没有涉及"在何时写的"这一动作,所以应该用一般现在时而不用过去时。

【改正后的句子】 **The preface to this book is well written.**

**例 304**

【汉语原句】 我们获得了 86.3% 的分辨率。

【英语错句】 We get 86.3% recognition rate.

【错误分析】 (1)对于"获得"这一含义的表达不少读者喜欢用"get",而英美人一般用"obtain",而且在此最好用现在完成时。(2)"分辨率"在此是一个可数名词,因此在其前面一定要有冠词。(3)"86.3%"不能直接放在名词前作定语,在英语中一般要用"of"短语来引出其前面名词的同位语。又如:

This battery can supply a current **of 4 mA**. (这里也可写成"a 4-mA current")

这个电池能提供 4 毫安的电流。

【改正后的句子】 **We have obtained a recognition rate of 86.3%.**

**例 305**

【汉语原句】 本文讨论了在半导体器件生产中所出现的非正态分布参数。

【英语错句】 This paper discussed the non-normal distribution parameter appeared in semiconductor device production.

【错误分析】 (1)表示"本文"作主语时(不论采用主动句还是被动句),其谓语动词均用一般现在时,所以应该把"discussed"改成"discusses"。(2)由于"appear"属于不及物动词,所以不能用它的过去分词作定语,而只能用现在分词作定语,因此应该把"appeared"改成"appearing"。(3)最好把"semiconductor device production"改成"the production of semiconductor devices"。

【改正后的句子】 **This paper discusses the non-normal distribution parameter appearing in the production of semiconductor devices.**

**例 306**

【汉语原句】 我们获得了具有 P 个顶点的 2-重自补图的数目是

$$a_p = 2(S_p^{(2)} ; 1, 3, 1, 3, \cdots)$$

【英语错句】 We obtain that the number of 2-edges self-complementary graphs with p vertexes is that

$$a_p = 2(S_p^{(2)} ; 1, 3, 1, 3, \cdots)$$

【错误分析】 (1)"获得了"最好用完成时态表示。(2)"obtain"是不能跟有宾语从句的,应在"that"前加一个名词"the result",这样"that"引导修饰"result"的同位语从句,这里也可以采用被动句的形式写成:"The result has been obtained that …"。(3)"2-edges"是作定语用的,应该使用单数形式,又如"a six-volt battery"(一只六伏特的电池)。(4)在同位语从句中的主语是"the number",所以其表语只能是一个名词(或形容词)而不能存在一个表语从句,所以句中"is"后的"that"应该去掉,这样那个表达式就作表语了。

【改正后的句子】 We have obtained the result that the number of 2-edge self-complementary graphs with p vertexes is

$$a_p = 2(S_p^{(2)} ; 1, 3, 1, 3, \cdots)$$

例 307

【汉语原句】 本文首先分析了量化误差会降低信号功率和距离分辨率。

【英语错句】 This paper first analyzed quantization error will reduce signal power and range resolution.

【错误分析】 (1)论文"分析"、"讨论"、"提出"等等均应该用一般现在时而不用一般过去时。(2)动词"analyze"后面是不能跟宾语从句的,本句中应该在"analyzes"后加上"the phenomenon[fact] that"。(3)由于"error"是可数名词单数,所以要在"quantization error"前加上冠词。

【改正后的句子】 This paper first analyzes the phenomenon/fact that the quantization error will reduce signal power and range resolution.

## IX、倒装、强调、虚拟语气问题

例 308

【汉语原句】 密码学意义上强的序列不仅应该具有足够高的线性复杂度,而且当少量比特发生变化时不会引起线性复杂度的急剧下降。

【英语错句】 Not only cryptographically strong sequences should have a large linear complexity, but also the change of a few terms should not cause a significant decrease of the linear complexity.

【错误分析】（1）由于本句以"not only"开头，所以一定要发生部分倒装现象，因此应该把句中的第一个"should"紧放在"Not only"后面。在英语中，需要部分倒装的常见情况如下：

① 当句子以"only＋状语（副词；介词短语；状语从句）"开头时必须发生部分倒装（其倒装方式与一般疑问句构成类同，即把助动词或情态动词放在主语前）。例如：

Most digital circuits talk to other digital circuits. Only rarely **must** a gate talk to the outside world.

大多数数字电路是与别的数字电路对接的。只有在极个别情况下门电路要与外界发生联系。

Only under a matched condition **is** there a maximum output.

只有在匹配条件下才能获得最大的输出。

Only if the force is constant and in the same direction as the displacement, **is** it true that "work equals force times distance."

只有在力是恒定的并且处于与位移同一方向时，"功等于力乘以距离"才成立。

② 当句子以"修饰谓语的否定性副词或否定性介词短语"开头时，句子要发生部分倒装，其倒装方法也与一般疑问句构成类同。常见的这类否定性副词和否定性介词短语有 never，hardly，scarcely，seldom，rarely，little，not only，not always，not until，nowhere，by no means，in no way，at no time，in no case，on no account，under［in］no conditions，under［in］no circumstances，on no condition，in no event，during no portion of …，at no other time in our history，at neither end of …等。例如：

Nowhere in the definition of computer viruses **is** there any mention of nonprompted, secret operations, of destructive actions, or of spreading across multiple computer installations.

在定义计算机病毒时，任何地方都没有提到无提示的秘密操作、破坏性的行为或者在多台计算机设施间传播。

Not always **does** the addition or removal of heat to or from a sample of matter lead to a change in its temperature.

把热量加给某一物质或从中取走热量，并不总会导致其温度的变化。

By no means **do** positive charges flow in a wire.

正电荷是绝不会在导线中流动的。

At no other time in our history, **has** success depended so heavily on intelligence and information.

在我们的历史上从来也没有像现在这样，成功如此大量地依赖于情报和信息。

③ 句子以"so(也)；neither 或 nor(也不)"开头时必须部分倒装，其倒装方式同样与一般疑问句构成类同。例如：

Two electrons will be repelled from each other and so **will** two nuclei.

两个电子相互排斥，两个原子核也是如此。

In the absence of friction, the driving wheel would not run the belt, neither/nor **would** the belt run the wheel to be driven.

若没有摩擦，主动轮就带不动皮带，而皮带也带不动从动轮。

④ 某些状语从句需要采用部分倒装句型，在科技文中常见的有：

ⅰ. 当"as"引导让步状语从句时（即："as"意为"虽然"时；有时"though"引出的让步状语从句也可发生部分倒装现象。）

Small **as** atoms are, electrons are still smaller.

尽管原子很小，但电子更小。

Simple **though** these crystal sets were, a fair degree of skill was needed in the adjustment of the "cat's whisker."

这些矿石收音机虽然很简单，但在调整"猫须"时需要相当技巧的。

ⅱ. 在"越……，越……"句型中

**The larger** the current is, **the greater** the power will be.

电流越大，功率就越大。

ⅲ. 虚拟语气句型中的"if"省略时，从句要发生部分倒装现象

We could not live on earth **were** there no air.

若没有空气，我们就不能在地球上生存。

**Should** anything abnormal happen, les us know as soon as possible.

万一发生什么不正常情况，请立即通知我们。

ⅳ. 让步状语从句连接词"whether"省去时，动词"be"要用原形且倒置在从句主语前

The chemical composition of water is $H_2O$, **be** it solid, liquid, or water vapor.

水的化学成分总是 $H_2O$，不论它是固体、液体还是水蒸气。

ⅴ. 在由"than"和"as"引导的比较状语从句及由"as"引导的方式状语从句或定语从句中，为了加强语气时也会发生部分倒装现象

This computer is much smaller <u>than</u> **is** a typewriter.

这台计算机比一台打字机小得多。

The flow of current meets some opposition just <u>as</u> **does** the flow of water.

电流的流动就像水的流动一样，会遇到某些阻力。

The characteristics of these tuned lines are calculated in the same way <u>as</u> **are** the characteristics of transmission lines.

这些调谐线的特性的计算方法，与传输线特性的计算方法相同。

⑤ 句型：

"介词短语或'there'＋动词的被动形式［或形容词(作表语)］＋主语［或宾语］"

In many modern laboratories **may be found** <u>electronic computers</u> which can make the complicated computations in a matter of minutes.

在许多现代的实验室里，人们可以发现能在几分钟左右的时间内进行极为复杂计算的电子计算机。

Fig. 1－5 shows a block on which **are exerted** <u>two forces $F_1$ and $F_2$</u>.

图 1－5 画出了在一木块上加有两个力 $F_1$ 和 $F_2$ 的情况。(倒装发生在定语从句中)

To each point of $S$ let there **be assigned** <u>a definite number</u>，$n$.

设对 $S$ 的每一点，赋予一个确定的数，$n$。

There **has been developed** <u>a voltage</u> across $C_2$.

在 $C_2$ 两端产生了一个电压。

There **is present** <u>a magnetic field</u> which varies with time.

(现在)存在有一个随时间变化的磁场。

There **are accessible** to you，in libraries，<u>any number of books</u> about different kinds of communication equipment.

在各图书馆里，你可以借阅到许多有关各种通讯设备的书籍。

⑥ 某些需要强调的成分放在主语前。例如：

**This property** we call inertia.

这一性质我们称之为惯性。(强调了宾语)

**Certain** it is that all essential processes of plant growth and development occur in water.

确定无疑的是，植物生长和发育的一切关键过程都是发生在水里的。(强调了主句中的表语)

**Of one thing** we can be sure：if the charges are at rest，there is no electric field at any point within the metal ball.

有一点我们是可以肯定的：如果电荷处于静止状态，则在金属球内部的任何一点上均没有电场。（强调了主句中的状语）

⑦ 把较短的宾语补足语或状语放在宾语前。例如：

The power economy of the transistor makes **reliable** such control equipment.

由于晶体管的功耗低，所以使得这种控制设备很可靠。

Only when we exert **on an object** an upward force sufficient in magnitude can we lift it.

只有当我们对一物体施加一个足够大的向上力时，才能把它提起来。

（2）句尾"linear complexity"前的定冠词"the"应该去掉。（3）在名词 increase，decrease，rise，fall，reduction 等后面，多数英美人喜欢用"in"，当然"of"并没有错。

**【改正后的句子】** Not only should cryptographically strong sequences have a large linear complexity，but also the change of a few terms should not cause a significant decrease in linear complexity.

**例 309**

**【汉语原句】** 直到发明了显微镜，对细菌的真正研究才成为可能了。

**【英语错句】** Until the invention of the microscope the real study of bacteria became possible.

**【错误分析】** （1）"until"这词本意为"直到……（之前）"，所以与汉语所表达的意思不符，要表示"直到……（之后）"一定要使用"not until"。（2）在句首存在有修饰句子谓语的否定性副词或否定性介词短语（如 not only，not until，hardly，little，seldom，never，scarcely，rarely，neither，nor，by no means，in no way，at no time，in no case，on no account，under no conditions，under no circumstances，on no condition 等）时，句子应发生部分倒装（其方式与构成一般疑问句相同），所以本句因"not until"开头而要发生部分倒装，在句子主语"the real study"前加助动词"do"的过去式"did"，并且将"became"变成动词原形"become"。

**【改正后的句子】** Not until the invention of the microscope did the real study of bacteria become possible.

**例 310**

**【汉语原句】** 然后出现了盲速问题。

【英语错句】　Then a problem arising, blind speed.

【错误分析】　(1)这个句子没有谓语,所以应该把"arising"改成"arises"。(2)"blind speed"实际上是"problem"的同位语,因此应该用"of"引出这个同位语,变成"the problem of blind speed"。(3)本句最好使用倒装句,即把谓语动词放在主语前。

【改正后的句子】　**Then arises the problem of blind speed.**

例 311

【汉语原句】　我们发现汇聚延迟存在门限值,只有当延迟超过门限值时,它才会严重地减少 TCP 吞吐量。

【英语错句】　This paper finds that there is a threshold of the assembly delay, only when the delay is larger than the threshold, the delay will severely reduce TCP throughput.

【错误分析】　(1)应该把"This paper finds"改为"It is found",因为论文不会发现什么的。(2)在"only"前应该加上"and that"。(3)由于第二个从句中是以"only+when 从句"开头的,所以后面应该发生部分倒装。(4)在"TCP throughput"前应该加定冠词。

【改正后的句子】　**It is found that there is a threshold of the assembly delay, and that only when the delay is larger than the threshold, will the delay severely reduce the TCP throughput.**

例 312

【汉语原句】　支承矢量机(SVM)要求核函数必须满足严格的 Mercer 条件。

【英语错句】　Support Vector Machines (SVMs) require that kernel functions must satisfy the rigorous Mercer's condition.

【错误分析】　由于在主句中有及物动词"require",所以根据语法规定,在宾语从句中(主句为被动语态时则在主语从句中)一定要使用虚拟语气,按照美国式用法,只要把"must"去掉就可以了。在科技文中,需要使用虚拟语气的常见的及物动词有:require(要求),desire(希望),suggest(建议),recommend(建议),propose(建议),necessitate(需要),demand(要求)等。

【改正后的句子】　**Support Vector Machines (SVMs) require that kernel functions satisfy the rigorous Mercer's condition.**

**例 313**

【汉语原句】　如果采用 ANN 的普通算法，就会出现过拟合问题。

【英语错句】　If the common algorithms of ANNs are adopted，it will result overfitting problem.

【错误分析】　(1) 本句的主句部分是典型的中文式英语句子，原作者把"result"看成是及物动词了。(2) 由于"problem"是可数名词单数，所以其最前面应该有冠词(在此应该用"an")。(3) 根据英美人的习惯，由于本句所述的情况是不愿看到的，因此作为告诫，本句最好使用虚拟语气。

【改正后的句子】　**An overfitting problem would result（或：arise）if the common algorithms of ANNs were adopted.**

**例 314**

【汉语原句】　对于 256 字节的信息来说，如果 $R=32$，则需要多达 64 个 CRC 计算器。

【英语错句】　For 256 bytes，there needed as many as 64 CRC calculators if $R=32$.

【错误分析】　(1) 在"256 bytes"之前应该加上"a message of"，因为它是一个信息。(2) 由于"need"在此用作为及物动词，所以不能采用引导词"there"开头的句型，应该改成"as many as 64 CRC calculators would be required")。(3) 本句用虚拟语气为好。

【改正后的句子】　**For a message of 256 bytes，as many as 64 CRC calculators would be required if $R=32$.**

# X、"with"结构及否定

**例 315**

【汉语原句】　最后对某加固计算机机箱在强迫风冷下的热特性进行了数值计算，获得了一些对电子设备热设计有价值的结果。

【英语错句】　In the end，the thermal characteristics of a certain reenforced computer case under the forced air convection are calculated. Some valuable results for the thermal design of electronic equipment are obtained.

【错误分析】　(1) 本句中开头的"In the end"是错误的，在此应该用"Finally"一词。(2) 这个汉语句原本是一个句子，逗号前后的逻辑关系很紧密，所以英语最好也用一个句子来表达，否则句子就显得松散。在此把第一个句子的句号改成逗号，而在"some"前加上"with"，"some"的第一个字母应小写，同时把"obtained"前的"are"去掉，这样就变成了一个"with 结构"，对前一部分作了附加说明。我们知道，科技写作的句子是很严格的，所以一般比较冗长而又紧凑，其常用的方法有以下三种：

① 采用"with"短语，分词短语，形容词短语等，例如：

An initial analysis of the device parameters has been made **with satisfactory results**.

我们对器件参数进行了初步分析，其结果令人满意。

（本句中用了一个"with"短语来表示汉语句子中逗号后的那部分内容。）

Silver is the best conductor，**followed by copper**.

银是最好的导体，其次是铜。

（本句中用了一个分词短语来表示汉语句子中逗号后的那部分内容。）

**Small in size and low in price**，this device is warmly received by users.

由于该设备体积小、价格低，所以很受用户欢迎。

（本句中用了一个形容词短语来表示汉语句中逗号前的那一部分内容。）

② 采用"with 结构"（这最常见）、"分词独立结构"等，例如：

A brief introduction to the principles of a computer is given, **with emphasis on the design of software**.

本文对计算机原理作了简要的介绍，重点放在软件的设计上。

（本句中用了一个"with 结构"来表示汉语句中逗号后的那一部分内容。）

An electron is about as large as a nucleus, **its diameter being about $10^{-12}$ cm**.

电子与原子核的大小大致相同，其直径大约为 $10^{-12}$ 厘米。

（本句中用了一个分词独立结构来表示汉语句中逗号后的那一部分内容。）

**With its base grounded**，Q4 is a very high impedance.

在其基极接地的情况下，Q4 是一个很高的阻抗。

（本句的汉语句中逗号前的那部分内容是用一个"with 结构"来表示的。）

③ 采用定语从句等来表示，例如：

Semiconductors are very sensitive to light and heat, **both of which have a great effect on their conductivity**.

半导体对光和热很敏感，这两者对其导电率影响很大。

（本句中用了一个定语从句来表示汉语句中逗号后的那一部分内容，因为汉语句中逗号前后的两部分内容中可找到共同的东西，即"这两者"，就是指

"光和热"，因此可用关系代词了。）

A new method for solving this problem is presented，**which is simple and practicable.**

本文提出了解决这一问题的一种新方法，这种方法简单而切实可行。

（在汉语原句中"方法"是逗号前后的两部分内容所共有的，所以可以用定语从句来表示逗号后的那部分内容。）

【改正后的句子】　**Finally，the thermal characteristics of a certain reinforced computer case under the forced air convection are calculated，with some valuable results for the thermal design of electronic equipment obtained.**

### 例 316

【汉语原句】　信道模型为 WSSUS，而 $L=16$。

【英语错句】　Channel model is WSSUS，and $L=16$.

【错误分析】　（1）在主语和表语前应该加定冠词。（2）"$L=16$"应该用"with 结构"来表示。

【改正后的句子】　**The channel model is the WSSUS with $L=16$.**

### 例 317

【汉语原句】　$\alpha$ 是调整平均功率的参数，本文中 $\alpha=1.4$。

【英语错句】　$\alpha$ is the parameter to adjust the average power，in this paper $\alpha=1.4$.

【错误分析】　（1）在"parameter"后一般跟"for 短语"。（2）在逗号后的那部分应该用"with 结构"来表示。

【改正后的句子】　**$\alpha$ is the parameter for adjusting the average power，with $\alpha=1.4$ in this paper.**

### 例 318

【汉语原句】　我们把 $A$ 看作为一个频率因子阵，矩阵行数 $M$ 取 64、列数取 $N$，$N$ 等于滤波器的阶数。

【英语错句】　We consider $A$ as a frequency factor matrix with 64-row and $N$-column，and $N$ equals to the order of the filter.

【错误分析】　（1）"64-row"应该为"64 rows"；"$N$-column"应该为"$N$ columns"。（2）逗号后"and"及其后面的部分应该用一个"with 结构"来表示，使句子紧凑。要注意："等于"作谓语时，应该表示成及物动词"equals"或"be+

形容词短语",即"is equal to",而绝不能用不伦不类的"equals to"。

**【改正后的句子】** We consider A as a frequency factor matrix with 64 rows and N columns, with N equal to the order of the filter.

#### 例 319

**【汉语原句】** 理论分析和实验结果表明,VGSR 算法不仅消耗能量低,而且能很好地适应网络规模的变化,其性能优于现有的其他算法。

**【英语错句】** The theoretic analysis and experiment results show that VGSR algorithm not only is low energy cost, but also scales well with the network size, the performance of which is superior to other existing algorithms.

**【错误分析】** (1)"理论分析和实验结果"应该写成"Theoretical analysis and experimental results"。(2)在"VGSR algorithm"之前应该加定冠词。(3)为了与后面内容匹配,所以"不仅消耗能量低"要写成"not only consumes a small amount of energy"(否则可以表示为"is not only low in energy consumption")。(4)采用"the performance of which"表示的话,"which"指代不清楚,因此应该用一个"with 结构"。(5)在"other existing algorithms"前应该加上"that of",否则比较对象不一致。

**【改正后的句子】** Theoretical analysis and experimental results show that the VGSR algorithm not only consumes a small amount of energy, but also scales well with the network size, with its performance superior to that of other existing algorithms(或:other algorithms available)。

#### 例 320

**【汉语原句】** $x$ 的所有值均不能满足这一方程。

**【英语错句】** All values of $x$ can't satisfy this equation.

**【错误分析】** (1)本句存在一个概念上的严重错误,主要是汉语与英语在表达上的差异引起的,读者由于受到了汉语的影响而经常产生这样的错误,也就是在英语中如何来表示全否定和部分否定的问题。若要表示全否定,英语中通常要使用 no、none、neither 这种词来表示。例如:

**None** of the problems can be solved at once.

所有这些问题均不能立即得到解决。

**No** textbooks available (或:**None** of the textbooks available) have mentioned this phenomenon.

现有的教科书<u>均没有</u>提到这一现象。

**Neither** of the devices is good in quality.

这两台设备的质量<u>都不</u>好。

而若使用"every，all，both"等与"not"处在同一句中时则一般表示部分否定，应该译成"并非都"之意（相当于"not every，not all，not both"）。例如：

**All** these values are **not** correct.

这些值<u>并非都</u>正确。

The velocity of a viscous fluid flowing through a tube is **not** the same at **all** points of a cross section.

流过管子的黏性流体的速度在横截面的各点上<u>并非都</u>是相同的。

（2）句中使用了"can't"这种形式，前面讲过，根据英美科技人员的写作习惯，一般是不用这种缩略形式的，要求写成"cannot"或"can not"。

【改正后的句子】　**No values of *x*（或：None of the values of *x*）can satisfy this equation.**

**例 321**

【汉语原句】　这个电压几乎是测不出的。

【英语错句】　This voltage almost can not be measured.

【错误分析】　句中受到汉语"几乎不能"的影响而用了"almost"加上"not"，实际上英美人是用"can hardly"来表示的。

【改正后的句子】　**This voltage can hardly be measured.**

# Ⅺ、两个介词共用一个介词宾语

**例 322**

【汉语原句】　一种对象/关系映射框架的研究和分析

【英语错句】　Research and Application about an Object/Relation Mapping Framework

【错误分析】　这是一篇论文的标题，其中的主要错误是"research"与"application"后面跟的介词是不同的。在英语中，许多名词后跟"of"来表示"的"的含义，但在某些名词后则是不能用"of"的，这里的"research"就是其中之一，在它后面可以跟"on""into"等（最常见的是跟"on"）；而在"application"后则是跟"of"的，这样就出现了"两个介词共用一个介词宾语"的情况，这在科技文中是经常遇到的。又如："必须求出这个电阻上的电压和电流。"应该写成"It is necessary to find out the voltage **across** and the current **through** this resistor.""这取决于离电台的距离和电台的方向"应写成"This depends on the distance **from** and the direction **of** the radio station.""本文讨论这类规划问题的性质及其全局算法。"应写成"This paper describes the properties **of** and a global algorithm **for** this type of programming problem.""这位教授在研究和应用沃氏函数方面起了积极的作用。"应写成"This professor has played an active role in the research **on** and the application **of** the Walsh functions."

**【改正后的句子】**　Research on and the Application of an Object/Relation Mapping Framework

**例 323**

**【汉语原句】**　同时给出了三轴稳定跟踪的数学描述及全补偿条件。

**【英语错句】**　The mathematical description and the full-compensation conditions of the 3-axes steady-tracking principle are also given.

**【错误分析】**　(1)"description"和"condition"后面所需跟的介词是不同的，在"description"后应该用"of"来引出它的逻辑宾语，而"condition"后一般要用"for"来表示"……的条件"。(2)" 数字＋名词"作定语时其中的名词必须用单数形式，又如"一个 6 伏的电池"应译成"a 6-volt battery"。所以这里的"3-axes"应改成"3-axis"。

**【改正后的句子】**　The mathematical description of and the full-compensation conditions for the 3-axis steady-tracking principle are also given.

**例 324**

**【汉语原句】**　本文讨论了三维多通道稳态流场和温度场的数值计算与实验。

**【英语错句】**　Numerical computation and experiment for the steady-state flow and temperature fields in three dimensional multi-channels are discussed in this paper.

**【错误分析】**　(1)英语中大多数由及物动词演变来的名词应该用"of"引出它逻辑上的宾语，所以在名词"computation"之后应加上"of"；而"experiment"后应跟介词"on"而不是"for"，同时在上述两个名词前应分别加上定冠词"the"。(2)在论文的文摘的第一句中不必加上"in this paper"，这是多余的。

**【改正后的句子】**　The numerical computation of and the experiment on the

**steady-state flow and temperature fields in three dimensional multi-channels are discussed.**

### 例 325

【汉语原句】　本文论述了该系统的设计思路与结构，深入探讨了模型检验。

【英语错句】　This paper discusses designing ways and structure of the system and deeply discussed model testing.

【错误分析】　(1)"……的设计思路"可表示成"the design guideline for …"。(2)在"structure"前应该加一个定冠词"the"。(3)为了使句子结构紧凑，本句"and…"这一部分可以用一个 with 短语来表示，即写成"with a deep inquiry into model testing"。

【改正后的句子】　**This paper discusses the design guideline for and the structure of the system，with a deep inquiry into model testing.**

### 例 326

【汉语原句】　建立了 SiGe HBT 直流特性模型及物理意义清晰的各电流解析表达式。

【英语错句】　The SiGe HBT DC models and the analytic expressions of various currents with clear physical meaning are established.

【错误分析】　(1)两个名词需跟有不同的介词而不是只用一个"of"，这叫"两个介词共用一个介词宾语"，在"expression"后英美人通常使用介词"for"，因此应该把本句的主语改写成"The SiGe HBT DC models **of** and the analytic expressions **for** various current …"。(2)由于电流有多个，所以"meaning"应该用复数形式。

【改正后的句子】　**SiGe HBT DC models of and analytic expressions for various currents with clear physical meanings are established.**

### 例 327

【汉语原句】　本文对模糊熵的理论研究与实际应用均有重要的意义。

【英语错句】　This paper contributes both to the theoretical research and the application of fuzzy entropy.

【错误分析】　(1)由于"both…and"为并列连词，所以它连接的东西一定要等列，其改正的方法有两种：一是在"the application"前加上介词"to"；一是把"both"放在介词"to"之后。（2）由于"research"与"application"后跟的介词是不同的，前者要跟"on"而后者是跟"of"来引出各自的逻辑宾语，所以在此可使用两个介词共用一个介词宾语的句型，即："the theoretical research **on** and the application **of** fuzzy entropy"。（3）动词应该用将来时，并且在"contribute"后面加上副词"significantly"。不过最好使用"This paper is of great significance to both …"。

【改正后的句子】 **This paper will be of great significance to both the theoretical research on and the application of fuzzy entropy.**

## XII、几个固定句型及句子成分的分隔

例 328

【汉语原句】 所谓曲线族，指的是能满足给定条件的一组特殊的曲线。

【英语错句】 The so-called family of curves refers to a special set of curves which satisfy given conditions.

【错误分析】 这个句子从语法上来说并没有错，而且科技概念也是对的。主要是每当表达本汉语句所述的含义时，实际上它是对某事的一种定义，我们称之为"定义句"的一种，这时最好使用"By A is meant B"（所谓 A 指的是 B）这一句型，或用它的主动形式"By A we mean B"（所谓 A 我们指的是 B）。又如：

**By linear operation is meant the ability** of an amplifier to amplify a signal with little or no distortion.

所谓线性工作，指的是放大器能以很小的失真或无失真地放大信号的能力。

**By elasticity is meant the tendency** of an object to return to its original state after being deformed.

所谓弹性，指的是一个物体在形变后能回到其原来状态的趋势。

注意：形容词"so-called（所谓的）"在科技文中很少使用，它在政论文章中主要表示贬义。在科技文中若要表示"所谓的；人们（通常）所说的［所称的］"的含义时，一般使用表示客观情况的"what 从句"，其句型如下（这里以被动形式为例）：

**what is called**［termed；named；described as；known as；referred to as；spoken of as；… ］＋名词（在从句中作补足语）

例如：

**What we call** a robot is no more than a special kind of electronic equipment.

我们所谓的[所说的]机器人，只不过是一种特殊的电子设备。

Late in 1947，they discovered **what was later to be named** the transistor effect.

在 1947 年后期，他们发现了后来人们所称谓的"晶体管"效应。

Magnitude，direction and place of application are **what we describe as** the elements of a force.

大小，方向，作用点就是我们所说的力的三要素。

Sending a signal from one place to another is **what is known as** transmission.

把信号从某地发送到另一处就是所谓的传输。

The reflected portion of waves forms **what is referred to as** the echo signal.

反射回来的那部分波形成了所谓的回波信号。

【改正后的句子】　**By a family of curves is meant a specified set of curves which satisfy given conditions.**

**例 329**

【汉语原句】　与普通的排列相比，这种新颖排列的优点是标度小、对高速处理的适应性好。

【英语错句】　Compared with the ordinary configuration，the advantages of this novel configuration are small in scale and good in adaptation to higher-rate processing.

【错误分析】　(1) 有名词"advantage"时，表示"与……相比"的含义应该用介词"over"来表示。(2) 本句的主语是抽象名词"advantages"，所以表语应该用名词短语来与它等同而不能用形容词短语，只要把句中两个"in"去掉即可，这是一些读者常犯的一个错误。不过英美科技人员经常喜欢用"主-谓-宾"结构来表示我们汉语中的"主表"结构，常见的谓语动词是"have"或其他有关动词。例如：

这根导线长 10 米。

This wire **has** a length of ten meters.

（等效于 The length of this wire is ten meters.）

这些钢板平均厚五厘米。

These steel plates **have** an average thickness of five centimeters.

（等效于 The average thickness of these steel plates is five centimeters.）

这个装置的结构比那个复杂。

This device **has** a more complicated structure than that one.

（等效于 The structure of this device is more complicated than that of that one.）

锡的熔点没有铅的高。

Tin does not **have** as high a melting point as lead does.

（等效于 The melting point of tin is not as high as that of lead.）

那类天线的优点是结构简单，效率高。

That type of antenna **has** the advantages of simple structure and high efficiency.

银具有的自由电子比铁多。

Silver **possesses** more free electrons than iron does.

这种方法所需的运算量远小于那种方法。

This technique **requires** a far smaller amount of operation than that one.

你使用的结构应尽可能的简单。

You should **use** as simple a structure as possible.

这个设备消耗的能量比那个少得多。

This device **consumes** much less energy than that one does.

**【改正后的句子】** **This novel configuration has the advantages over the ordinary configuration of small scale and good adaptation to higher-rate processing. （或：The advantages of this configuration over the ordinary one are small scale and good adaptation to higher-rate processing.）**

要注意：这里的"of small scale and good adaptation to higher-rate processing"是修饰"the advantages"的，这叫做"句子成分的分割"。在科技文中，关于这种"分割"现象，读者应着重注意以下几点：

（1）"主语与其修饰语的分隔"，即"主语＋谓语＋主语修饰语"句型。

这一句型的主要目的是为了防止句子发生"头重脚轻"现象，其主要形式如下：

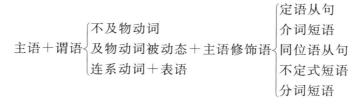

这里的"表语"以形容词为多见。例如：

**The evidence is conclusive that** electric charge is not something that can be divided indefinitely.

有确凿的证据表明，电荷并不是可以无限加以分割的某种东西。

**The probability is 0. 99 that** a coherent spectral line will exceed a neighboring noise line.

一根相干谱线会超过其相邻的一根噪声谱线的可能性为0.99。

**No solution exists to** Equation (10) for sufficiently positive values of $v$.

若 $v$ 的正值足够大，则方程[10]无解。

**Opinions differ considerably as to** how descriptive chemistry should be covered in a general chemistry textbook.

关于在普通化学教科书中如何来讲解描述性化学的意见很不一致。

**Very wonderful changes in matter take place** before our eyes every day **to which** we pay little attention.

十分奇妙的物质变化天天都在我们的眼前发生着，可是我们却几乎没有注意到它们。

**No charges have ever been found of** smaller magnitudes than those of a proton or an electron.

至今尚未发现哪个电荷量比质子或电子所带的电量更小。

Now **the reasonable assumption can be made that** all ions of the same kind carry the same charge.

现在有理由认为：同类电子均带有相同的电荷。

In Chapter 8，**a design procedure** is developed **which** is the reverse process of the analytical technique.

第八章讲解设计步骤，它是分析方法的逆过程。

（2）一个名词后同时跟有两个修饰语（特别是第二个修饰语为"of-短语"）的情况。

① 一般情况，其形式为：

"被定的名词＋定语（1）（偶尔为状语）＋定语（2）"

（最常见）　　　　短语　　　　　　　　短语

（次常见）　　　　短语　　　　　　　　从句

　　　　　　　　　从句　　　　　　　　从句

　　　　　　　　　从句　　　　　　　　短语

例如：

It is now possible to find the Thevenin equivalent of the circuit **in Fig. 3** with no load by nodal analysis.

现在可以用节点分析法求出图 3 所示的、不带负载的电路的戴文宁等效电路。

This illustrates the coupling **between input and output terminals** inherent in transistors.

这表明了晶体管所固有的、输入和输出端之间的耦合。

Instability is the **tendency in certain systems** of a quantity associated with energy，such as current，to increase indefinitely in the absence of excitation.

不稳定性就是在某些系统中与能量有关的某个量，比如电流，在没有外部激励的情况下会无限增长的趋势。(注意：第二个定语是由"of"引出的不定式复合结构修饰"the tendency"。)

Collector modulation has the advantages **over base modulation** of better linearity，higher collector efficiency and higher power output per transistor.

集电极调制与基极调制相比的优点是：线性比较好、集电极效率比较高、每个晶体管的输出功率比较大。(注：第二个定语是由"of"引出"the advantages"的同位语。)

These basic virtues of personal computing have resulted in the growth **of what was a cottage industry in the early 1980**s to the multi-billion-dollar PC hardware and software industries of today.

由于个人计算的这些基本优点，促使八十年代初期的家庭工业成长为今天上百亿美元的 PC 硬件和软件工业。(注：第二个定语实际上是名词"the growth"的逻辑状语；另外，由于"what 从句"的谓语是"系表"结构，所以汉译时只要翻译其表语就行了。)

What lies behind this meeting is an increasing awareness **around the world** of the urgency of reducing global warming.

开这个会的背景是全世界越来越认识到了降低全球性变暖程度的紧迫性。(注：第二个定语实际上是名词"awareness"的逻辑宾语。)

In the preceding section，we established rules **to follow** which enabled us to systematically analyze circuits.

在上一节，我们确立了一些应遵循的规则，这些规则使得我们能对电路进行系统的分析。

This is what is meant by the ⟨statement⟩ **sometimes made** that "the pressure in a fluid acts in all directions."

这就是有时候我们说"流体中的压力是向四面八方作用"的含义。

There are a number of ⟨other striking examples⟩ **that can be given** which bear upon heat and temperature.

我们可以举出另外一些与热和温度有关的、引人注目的例子来。

There are ⟨a great many problems⟩ **which arise in the various fields of technology** which require for their solution methods beyond those available for algebra and trigonometry.

在技术的各个领域会出现许许多多的问题，而为了求解这些问题需要用到超出代数和三角所能提供的方法。

（注：在第二个定语从句中，"require"的宾语是"methods"，由于该宾语带有一个比较长的后置定语，所以把它放在状语"for their solution"["solution"与"their"之间存在"动宾关系"，整个介词短语译成"为了解它们"]之后了，可不要把"their solution methods"看成是一起的而译成"它们的解法"，如果是这样的话，那么及物动词"require"就没有宾语了，这是不允许的。）

Now we shall study the ⟨effects⟩ **that forces have** on motion.

现在我们将研究一下力对于运动的作用。

The ⟨opposition⟩ **that a conductor offers** tending to impede the transmission of electricity is called electrical resistance.

导体所提供的、趋于阻止导电的阻力被称为电阻。

② 两个特殊句型：

ⅰ."被定名词（是一个参数）＋介词短语＋of-短语（该参数的值）"

This assumption would lead to a **battery voltage** in the model of Fig. 7 of 840 mV.

这样假设后就可得到图 7 模型中的电池电压为 840 毫伏。

This analysis yields a **thermal resistance** from the ferrite to the coolant channel wall of approximately 36℃/W.

通过这一分析得到了从铁氧体到冷却剂通道壁的热阻近似为 36 摄氏度每瓦。

ⅱ."被定名词＋介词短语（该介词是与前面名词搭配使用的）＋of-短语"

要注意两点：

第一点：原来的搭配模式为"被定的名词＋of-短语＋介词短语（该介词是被定的名词所要求的）"，现在由于"of-短语"比较长而放在后面了，千万不要把这个"of-短语"看成是修饰其前面的介词短语中的介词宾语的。

第二点：如果被定的名词来自于不及物动词，则"of"后的介词宾语是该名词的逻辑主语；若被定的名词来自于及物动词，则"of"后的介词宾语为该名词的逻辑宾语，这时其逻辑主语可用放在"of 短语"后的"by 短语"来表示，而如果这个"by 短语"比较短的话，可以把"by 短语"放在"of 短语"之前，那么要特别小心不可把"of 短语"看成是修饰"by"的介词宾语的，这时的句型为"被定名词＋by-短语＋of-短语［＋其他的东西］"。例如：

An added advantage of this method is that it makes it possible for us to see the effect **on the overall waveform** of the absence of some of the constituents (for instance, the higher harmonics).

该方法的另外一个优点是使我们能看到缺少了某些分量［例如高次谐波］对于总波形的影响情况。（注：原来的搭配模式为"the effect of A on ［upon］ B"。）

This is realized by the provision **to the customer** of the means to support services.

其实现的方法是给用户提供支持服务的手段。（注：原来的搭配模式为"the provision of A to B"。）

The graphs in Fig. 4 summarize the variations **with frequency** of the resistance of a resistor, and of the reactances of an inductor and of a capacitor.

图 4 中的那些曲线概述了电阻器的电阻及电感器和电容器的电抗随频率的变化情况。（注：原来的搭配模式为"the variation of A with B"。）

What the forward bias achieves essentially is the injection **into the depletion layer** of electrons from the conduction band of the N-type material.

正向偏置的主要作用在于把来自 N 型材料导带的电子注入耗尽层中去。（注：原来的搭配模式为"the injection of A into B"。）

Evaporation from the soil and transpiration from vegetation are responsible for the direct return **to the atmosphere** of more than half the water that falls on the land.

土壤的蒸发和植物的蒸腾促使降在大地上的一半以上的水分直接返回到了大气层。（注：原来的搭配模式为"the return of A to B"。）

The instantaneous value of a quantity that varies sinusoidally with time is represented by the $\boxed{\text{projection}}$ **onto a horizontal axis** of a vector of length corresponding to the amplitude of the quantity and rotating counterclockwise with angular velocity ω.

　　随时间正弦变化的某个量的瞬时值是用一个矢量在水平轴上的投影值来表示的，该矢量的长度相应于该量的幅值，并且该矢量以角速度 ω 作反时针旋转。（注：原来的搭配模式为"the projection of A onto B"；另外，"of length corresponding to …"和"rotating …"这两个短语都是修饰"矢量"的。）

Pollution may be caused by the $\boxed{\text{release}}$ **by man** of completely new and often artificial substances into the environment.

　　污染可以由人们把崭新的、往往是人造的物质扔到周围环境中造成。（注：原来的搭配模式为"the release of A into B by C"，现在成为"the release by C of A into B"了。）

The $\boxed{\text{definition}}$ **by the standards bodies** of services and protocols for management applications has lagged behind their definition for business applications.

　　由制定标准的机构对应用于管理的服务和协议书所作的定义滞后于为商业应用而对它们的定义。（注：原来的搭配模式为"the definition of A by B"；另外，"definition"与"their"之间存在着"动宾关系"，所以应译成"定义它们；对它们的定义"。）

　　（3）其他情况。

Let us instead **use** for the upper limit on the integral the variable t.

　　而是让我们对积分上限使用变量 t。（注："use"的宾语被作状语的"for-短语"分隔开了。）

For the circuit above, let **the current $i(t)$** that is produced by the source be described by the function of time shown in Fig. 4.

　　对于上面的电路，设由电源所产生的电流用图 4 所示的时函数来描述。（注：本句中的宾语与宾语补足语被一个定语从句分隔开了。）

The material exhibited **unique properties** as it melted from a solid to a liquid state that could not be explained by the naïve understanding of matter at the time.

　　当该物质从固态熔化为液态时呈现了一些独特的性质，这些性质在当时是无法根据对物质的幼稚理解来解释的。（注：本句中的宾语与其定语从句被由

"as"引导的状语从句分隔开了。)

**例 330**

**【汉语原句】** 波能绕其通道上的障碍物的边缘弯曲行进，这一特性称为绕射。

**【英语错句】** Waves are able to bend around the edge of an obstacle in their path, this property is called diffraction.

**【错误分析】** 科技英语中在两个并列的单词、短语或分句之间应该用等立连词 and 来连接，而不少人在写作时却经常忘掉。不过，为了使科技文的句子严密而紧凑，本句后半部分可采用名词短语作为前面整句的同位语这一结构，其模式为(注意其汉译法)：

(句子)，
- a[an]＋形容词＋名词
- a[an]＋名词＋后置定语
  - 过去分词短语
  - 定语从句
  - 形容词短语
  - 同位语从句

例如：

The car noses up when it accelerates, **a familiar effect.**

小汽车加速时车头会向上抬起，这是大家熟悉的一种效应(或：这种效应是大家所熟悉的)。

Most metals may be deformed considerably beyond their elastic limits, **a property known as ductility.**

大多数金属形变的程度可以远远超过它的弹性限度，这一性质称为延性。

All forces fall into one or the other of these two classes, **a fact that will be found useful later.**

所有的力均可归属为这两类中的某一类，这一点以后会发现是很有用的。

**【改正后的句子】** Waves are able to bend around the edge of an obstacle in their path, a property called diffraction.

**例 331**

**【汉语原句】** 应该指出，这些信号在数值上没有很大的区别。

**【英语错句】** It should be pointed out that these signals haven't much magnitude difference.

【错误分析】　本句的不妥之处是在主语从句中，英美人常用以下两种句型来表示，一是写成"these signals do not differ greatly in magnitude"，一是写成"there is no great difference in magnitude between these signals"。

【改正后的句子】　**It should be pointed out that these signals do not differ greatly in magnitude.**

例 332

【汉语原句】　利用经典的振幅相关法来校准回波的包络会产生漂移的跳跃误差。

【英语错句】　There exist the drift and jump errors using the classical amplitude correlation method to align the envelope of the echoes.

【错误分析】　"利用……"这一部分从科技概念上看是一个条件，在这种情况下"会产生……误差"，所以我们可以采用"…errors will result if…is used to…"。另一种修改方法是把"using…"作为动名词短语作主语（因为这种主语往往可以表示一种条件），谓语动词用"result in…"或"lead to"等，即可以写成"Using the classical amplitude correlation method to align the envelope of the echoes will result in drift and jump errors."

【改正后的句子】　**Drift and jump errors will result if the classical amplitude correlation method is used to align the envelope of the echoes.**

例 333

【汉语原句】　这个系统与运动 BSC 有一些基本的差异。

【英语错句】　This system has some basic differences from the motion BSC.

【错误分析】　(1) 这个句子应该采用"there are … differences between A and B"结构。(2) 这里的"motion"应该改为"motional"。

【改正后的句子】　**There are some basic differences between this system and the motional BSC.**

例 334

【汉语原句】　本文首先讨论了这种信号的特点，然后讨论了它的产生。

【英语错句】　This paper first discusses the features of this signal, and then its generation is described.

【错误分析】　从语法上讲本句无错误之处，但由于这是一个并列复合句，一般来说两个并列句中的语态应该一致（有时时态也应一致），所以"its generation is described"应改成"describes its generation"。当然也可以把第一部分改成被动句而"and"后的那部分保持不变。

【改正后的句子】 **This paper first discusses the features of this signal, and then describes its generation.**（更好的版本为：**This paper begins with the discussion on the features of this signal, followed by the description of its generation.**）

**例 335**

【汉语原句】 最后将 XD 和传统大容量交换网络进行了性能比较，说明这是一种具有良好性能比和可扩展性的交换网络结构。

【英语错句】 Finally, the performance comparison of XD with traditional switching fabric of large capacity is given. It is shown that XD is a cost-effective and expandable switching fabric.

【错误分析】 (1) 这两个句子应该合并成一个长句子，这样句子结构就显得紧凑了，把句号变成逗号，把"it is shown"变成"which shows"。(2) "A 和 B 在……方面进行比较"最好表示成英美人喜欢的表达方式"a comparison in … between A and B"。(3) 与"comparison"搭配的动词一般是"make"。(4) "大容量"在此是定语，科技文中常用"large-capacity"作前置定语。(5) 从句子含意可知"XD"是一种交换网络结构，所以是可数名词，因此其前面应该有冠词（但根据观察，在"between A and B""from A to B""the variation of A with B"等情况下，"A"和"B"之前的冠词是可以省去的）。

【改正后的句子】 **Finally, a comparison in performance between XD and traditional large-capacity switching fabric is made, which shows that the XD is a cost-effective and expandable switching fabric.**

# 第三部分　句子松散等问题

**例 336**

【汉语原句】　模拟结果表明，所提出的 HBSD 算法比 MSD 算法来得好，且其复杂度约为 MSD 算法的 1/5。

【英语错句】　Simulation results show that the performance of the proposed HBSD algorithm is better than the MSD algorithm，and its complexity is about one-fifth of the MSD algorithm.

【错误分析】　(1) 在"than"后面要加"that"，否则比较对象不一致。(2) 在"and"后应该加第二个宾语从句的引导词"that"。(3) 在"of the MSD algorithm"前加代词"that"，也属于比较对象要一致。不过最好把"and…"改成一个"with 结构"，使句子显得紧凑。

【改正后的句子】　**Simulation results show that the performance of the proposed HBSD algorithm is better than that of the MSD algorithm，with its complexity being about one-fifth that of the MSD algorithm.**

**例 337**

【汉语原句】　该算法引入了自适应 SINR 门限以及"投票模型"，充分考虑了 LTE 网路中的小区类型。

【英语错句】　The SINR threshold and "vote model" are introduced in this algorithm. It fully takes into account the cell type in the LTE network.

【错误分析】　(1) 应把第二个句子改成一个非限制性定语从句，使句子显得紧凑。(2) "fully"应该改成"full"放在"account"前，这是英美人的习惯。又如"pay full attention to…"，"make full use of…"等。

【改正后的句子】　**The SINR threshold and "vote model" are introduced in this algorithm，which takes into full account the cell type in the LTE network.**

**例 338**

【汉语原句】　本文提出一种 FIB 多级映射的并行多流水路由查找架构 FMML，设计了动态路由映射算法。

【英语错句】　This paper presents a FIB multi-level mapping pararell multi-pipeline routing lookup architecture（FMML），and a dynamic routing mapping algorithm is designed.

【错误分析】　（1）在"FIB"前应该使用"an"。（2）把逗号后的一个分句改成一个"with 结构"使句子紧凑。

【改正后的句子】　**This paper presents an FIB multi-level mapping pararell multi-pipeline routing lookup architecture（FMML），with a dynamic routing mapping algorithm designed.**

例 339

【汉语原句】　（本文）提出了一种基于多模谐振器的小型三通带微带滤波器。该滤波器由一个模振荡器和一对馈线构成。

【英语错句】　A compact tri-band microstrip filter based on a MMR（multiple mode resonator）is presented. The filter consists of a MMR and a pair of feedlines.

【错误分析】　（1）在"MMR"前应该使用"an"。（2）第二个句子应该改成一个非限制性定语从句。

【改正后的句子】　**A compact tri-band microstrip filter based on an MMR（multiple mode resonator）is presented，which consists of an MMR and a pair of feedlines.**

例 340

【汉语原句】　文中比较了 STM 和 ATM，表明 ATM 将是 BISDN 的基础。

【英语错句】　STM and ATM are compared. It shows that ATM should be the ground of BISDN.

【错误分析】　（1）英语译文用了两个句子，显得松散，第二个句子应改成由"which"引导的非限制性定语从句。（2）"should"应改成"will"。（3）科技上的"基础"很少用"ground"一词的，应使用"basis"或"foundation"。

【改正后的句子】　**A comparison of STM with（或：and）ATM（或：A comparison between STM and ATM）is made，which shows that ATM will be the basis of BISDN.（甚至可以把它进一步精简成：A comparison between STM and ATM shows that ATM will be the basis of BISDN.）**

**例 341**

【汉语原句】　这个方法需要精确的定位，这就使得它的实现具有一定的难度。

【英语错句】　This method requires accurate location and this makes its realization have a little difficulty.

【错误分析】　(1)"have a little difficulty"属于中文式的英语句，这里只要使用"somewhat difficult"就可以了。(2)句中用了"and"显得比较松散，为使其前后两部分紧密联系在一起，我们可以使用"which"引导非限制性定语从句或使用分词短语表示结果状语。

【改正后的句子】　**This method requires accurate location，thus making（或：which makes）its realization somewhat difficult.**

**例 342**

【汉语原句】　我们可以由式(3)解 $\alpha_n$，从而得到 $U_\circ$。

【英语错句】　We can solve $\alpha_n$ by equation (3)，and thus obtain $U_\circ$.

【错误分析】　(1)看来原作者并不清楚"solve A for B"这一句型，意为"解 A 求出 B"，所以本句中应该使用"solve equation(3) for $\alpha_n$"。另外，多数英美人对"方程"的第一个字母用大写。(2)后面部分可以用分词短语作结果状语，这样使句子更紧凑了。

【改正后的句子】　**We can solve Equation (3) for $\alpha_n$，thus obtaining $U_\circ$.**

**例 343**

【汉语原句】　本文提出了一个基于离散对数的动态秘密分享方案，它能够检测欺诈者。

【英语错句】　In this paper，a sharing dynamic secret scheme based on discrete logarithms is proposed and it can detect cheaters.

【错误分析】　(1)句中"动态秘密分享"应该表示成"dynamic secret sharing"，又如"产油国"为"oil-producing countries"，"测温设备"为"temperature-measuring devices"等。(2)"and …"这一部分如果使用一个定语从句则句子就能显得紧凑了，即把它改成"which can（be used to）detect cheaters"。

【改正后的句子】　**In this paper，a dynamic secret sharing scheme based on discrete logarithms is proposed which can（be used to）detect cheaters.**

**例 344**

【汉语原句】　交流电可被转换成直流电，这一过程被叫做整流。

【英语错句】　AC can be turned into DC and this process is referred to as rectification.

【错误分析】　从语法角度讲本句没有错，但句子显得松散，最好把"and"变成一个逗号而把后面的东西用一个名词短语作其前面内容的同位语来表示（在"固定句型"部分已讲过），这样句子就得到了简化（从一个并列复合句变成了一个简单句）且紧凑，这一句子结构在科技英语中使用得很普遍。例如：

It may happen also that the cells of the blood donor are dissolved or go into solution, **a most dangerous condition.**

也可能会发生这样的情况：供血者的细胞被分解或溶解，这种情况是极其危险的。

Upon making use of this equation, the exact frequencies found in the hydrogen spectrum are obtained—**a remarkable achievement.**

利用了这个式子后就得到了在氢的频谱中所发现的那些频率，这是一个了不起的成就。

A portion of the output is returned to the input terminals, **a condition known as voltage feedback.**

输出电压的一部分被返回到了输入端，这种情况称为电压反馈。

The careful study of spectral lines shows that many of them actually consist of two or more separate lines that are close together, **something that the Bohr theory cannot account for.**

对谱线进行仔细的研究就会发现，它们中有许多实际上是由紧靠在一起的两根或多根分开的谱线组成的，这一点波尔理论是无法解释的。

From equation (1－2),

$$KE = \frac{I\omega^2}{2},$$

**a relation which is exactly analogous to**

$$KE = \frac{mV^2}{2}$$

**in linear motion.**

由式(1－2)得到 KE＝(1/2)$I\omega^2$，这一关系式与直线运动的 KE＝(1/2)$mV^2$ 完全类同。

【改正后的句子】　**AC can be turned into DC, a process referred to as** rectification.

### 例 345

【汉语原句】　本文分析了普通形式的交错 PRI 系列的固有性质,这种性质是:它们的次 PRI 能够形成一个算术级数。

【英语错句】　This paper analyzes the inherent property of a common style of stagger PRI trains. The property is that their sub-PRIs can form an arithmetical series.

【错误分析】　这两个句子在语法上没有错误,但显得松散,因此应该把它们合并成一个主从复合句,其做法是把第一句的句号去掉,同时把"The property is"也去掉即可,这样"that"就引导"the inherent property"的同位语从句了。

【改正后的句子】　**This paper analyzes the inherent property of a common style of stagger PRI trains that their sub-PRIs can form an arithmetical series.**

### 例 346

【汉语原句】　本文基于随机数的理论检验方法和经验检验方法对仿真过程中的随机发生器的选择作出建议。

【英语错句】　The paper gives suggestion for selecting random number generators in simulation. It is based on theoretical and empirical test methods for random numbers.

【错误分析】　(1)表示"提出(建议)"的动词常见的有"put forward, advance, make, offer"等,由于"建议(suggestion)"属于可数名词,所以其单数形式前必须要有冠词。(2)从汉语句来看,所述内容很紧凑,所以用英语来表示时最好用一个句子,否则显得松散,因此应该把中间的句号改成逗号,同时把"based on …"前面的"It is"去掉。(当然最好把"Based on…"放在句子的最前面。)

【改正后的句子】　**Based on the theoretical and empirical test methods for random numbers, this paper offers a suggestion for selecting random number generators in simulation.**

### 例 347

【汉语原句】　详细介绍了多描述分层编码方法,它对每一描述能产生基层和增强层,这使视频码流具有更强的抗误码性。

【英语错句】　The multiple description layered coding method is presented detailedly. The method generates base layer and enhancement layers for each description, which will make the video stream more robust.

【错误分析】 （1）虽然大词典里有"detailedly"一词，但实际上英美人在科技文章中是不用的，应该使用"in detail"。（2）由于"The method"就是前面提到的那个"方法"，为使句子紧凑，应该把它改成"which"，而把前面的句号改成逗号。（3）在"base layer"前应该加上定冠词"the"。（4）把"which will make"改成分词短语作结果状语"（thus）making …"。这样处理后，整个句子显得紧凑而精炼。

【改正后的句子】 **The multiple description layered coding method is presented in detail, which generates the base layer and enhancement layers for each description,（thus）making the video stream more robust.**

**例 348**

【汉语原句】 考虑到这些因素，提出了 N-ERA 算法，其特点是量化路径选择和合理的带宽调配。

【英语错句】 With these factors into consideration, an N-ERA is developed. It features quantized path selection and appropriate bandwidth allocation.

【错误分析】 （1）把"into consideration"改成"in mind"。（2）把句号改成逗号，并且把"It"改成"which"。

【改正后的句子】 **With these factors in mind, an N-ERA is developed, which features quantized path selection and appropriate bandwidth allocation.**

**例 349**

【汉语原句】 这个电荷与存在的其他电荷相互作用。

【英语错句】 This charge interacts with other present charges.

【错误分析】 句中"present"的位置放错了。这个形容词作前置定语时意义为"目前的"，而表示"存在"时一般要放在被修饰词的后面做后置定语。形容词作后置定语是中国人学习英语时不熟悉的内容之一。形容词作后置定语的主要情况如下：

① 在科技文中常见必须后置的形容词主要有五个："present"（意为"存在的"），"else"（在疑问代词和不定代词之后，意为"其他的，别的"），"what(so)ever"（在有否定词"no"等或"any"修饰的名词后，意为"任何的"），"involved"（意为"有关的；涉及的"）和"inclusive"（意为"首末包括在内的"）。例如：

The nucleus contains a total positive charge equal to the number of protons **present**.

原子核中包含的正电荷的总数与存在(于该原子核中)的质子数相等。

Everything **else** in this equation can be measured except $C^2$.

除了 $C^2$ 外，本式中的其他参数均能测得。

Up to this point，we have not made any approximations **whatsoever**.

至此我们并没有做任何的近似。(注：这里"any"已经把"whatsoever"的含义包括进去了。)

Eq. (1-7) can be used to determine any of the variables **involved**.

式(1-7)可以用来确定所涉及的任一变量。

In Chapters 9 to 14，**inclusive**，the detailed design of each of the components shown in Fig. 1-4 is treated.

在第 9 章到第 14 章(第 9 章和第 14 章包括在内)，将讲解图 1-4 所示的每个部件的详细设计。(注："inclusive"前的逗号可有可无。)

② 有些形容词常放在被修饰的名词后以加强语气，常见的有"available，obtainable，achievable，responsible，possible，usable，inclusive，total"等。例如：

This formula can be found in the physics books **available**.

这个公式可在现有的物理书中找到。

These are the smallest particles **obtainable**.

这些是所能获得的最小微粒。

③ 形容词只能放在由"some，every，any，no"与"thing，body，one"组成的不定代词之后。例如：

This book contains something **new**.

这本书里有些新内容。

Everything **electronic** will be done digitally.

所有的电子设备都将数字化。

④ 在由"and，or，both…and，either…or"等连接的两个形容词作定语时往往后置以加强语气。例如：

This book is a help to circuit designers **both new and old**.

本书对新老电路设计人员都是有帮助的。

Yet there are connections，**neither series nor parallel**，for which the above-mentioned phenomenon occurs.

然而还有一些连接方式也会出现上述现象的，它们既不是串联连接，也不是并联连接。(注：这里是一个非限制性定语。)

⑤ "副词或数量状语＋某些形容词"常作后置定语。例如：

Let us consider a current-carrying wire **10 meters long**.

我们来考虑一下一根<u>十米长的</u>载流导线。

⑥ 由"形容词＋介词短语（或动词不定式，或某些状语从句）"构成的形容词短语只能作后置定语。例如：

The electrons **able to move freely within a wire** play an extremely important role in the formation of electric current.

<u>在导线内能自由运动的</u>电子在形成电流方面起了极为重要的作用。

Infinity is a quantity **greater than any number**.

无穷大是<u>大于任何数的</u>一个量。

【改正后的句子】　**This charge interacts with other charges present.**

**例 350**

【汉语原句】　自动骨架提取法

【英语错句】　An Automatic Extraction Method of Skeleton

【错误分析】　（1）这是一篇论文的标题，由于受汉语的影响而出现了错误。"skeleton"是"提取"的宾语，所以这两个含意应该紧靠在一起。（2）在"method"后可以跟"for"短语、"of"短语或不定式短语，根据观察，它跟"for"短语的情况最为常见，所以在此采用"for the automatic extraction of［for automatically extracting］the skeleton"。

【改正后的句子】　**Method for the Automatic Extraction of the Skeleton**（凡论文标题开头的冠词一般省去）

**例 351**

【汉语原句】　本文构筑了一种适合于硬件实现的提升式 9－7 小波滤波器。

【英语错句】　This paper constructs a kind of lifting 9-7-tap wavelet filter for hardware implementation.

【错误分析】　（1）一般来说"本文"不会发出"构筑"这一动作，它通常可以与"论述；提出；讨论；分析"等动词连用。所以这里应该用"describes［presents］the construction of"来替代"constructs"。（2）"for"之前应该加上形容词"suitable"为好。

【改正后的句子】　**This paper describes the construction of a kind of lifting 9-7-tap wavelet filter suitable for hardware implementation.**

**例 352**

【汉语原句】 本文介绍了 IBM 个人计算机磁盘操作系统，也就是人们常说的 PC‒DOS。

【英语错句】 This paper introduces the IBM Personal Computer Disk Operating System, or PC‒DOS as it is commonly referred to as.

【错误分析】 本句中的错误是应该去掉句尾的"as"，因为这个介词与从句引导词"as"合二而一了，这种现象在英语中经常会见到。例如：

In this case the transmission line behaves **as** if it were short-circuited.

在这种情况下，该传输线的作用好像它是短路似的。（注：这个"as"是"behave as"与"as if"共用了。）

A text-book writer does not have available the material on which to base a reasonable conclusion as **to** whom the credit belongs.

一本教科书的作者不可能获得作出有关该感谢谁的合理结论的资料。（注：这里的介词"to"是"as to"和"belong to"共用了。）

【改正后的句子】 **This paper introduces the IBM Personal Computer Disk Operating System，or PC‒DOS as it is commonly referred to.**

**例 353**

【汉语原句】 本文注意到该系统具有永久磁铁的特点，提出一种比较简易的方法，解决其磁路的计算问题。

【英语错句】 This paper pays attention to having the characteristic of permanent ring magnet, and presents a kind of simpler method of solving problem of calculation for the magnetic circuit.

【错误分析】 （1）本句中的第一部分没有根据汉语句中的确切含义来翻译而是采用了逐字翻译法，应该把它写成"on the basis of the fact that this system has the property of a permanent ring magnet"。（2）第二部分中的"kind of"应该去掉。（3）"solving problem of"也要去掉。

【改正后的句子】 **On the basis of the fact that this system has the property of a permanent ring magnet，this paper presents a simpler method of calculation for the magnetic circuit.**

**例 354**

【汉语原句】 其理由可能是从时钟发生器到每个 PE 的距离不同之故。

【英语错句】 The reason for this may come from the different distances from the clock generator to each PE.

【错误分析】　本句主要是句子的主语与谓语在概念上是不匹配的，修改的方法有几种。该句可以改成"This may arise［result］from the different distances…"；也可以改成"This may be due to the different distances…"；还可以改成"The reason for this may be that the distances from the clock generator to each PE are different."

【改正后的句子】　**This may arise from the different distances between the clock generator and each PE.**

例 355

【汉语原句】　与 MEI 方法进行了对比，指出 MEI 方法中的第三假设值得商榷。

【英语错句】　After comparison with MEI method，it is shown that the third postulate of MEI method is questionable.

【错误分析】　在科技英语写作中，往往用表示动作的抽象名词（或其他名词）作主语来表示条件、时间、原因等，这样使句子结构更紧凑，并使概念表达得简洁明了。例如：

**A temperature rise of 50℃** would result in a change in Zener voltage of 3.75%.

如果温度升高 50℃就会导致齐纳电压改变 3.75%。

**Application of KVL to the collector-emitter loop** yields the following simple equation.

若把 KVL 应用于集电极－发射极回路就能获得以下的简单方程。

**The absence of atmosphere in space** would enable scientists to make pure drugs there.

由于在太空中不存在大气，所以能使科学家在那儿制造出纯净的药物。

**Too large a value of Q** would damage the device.

如果 Q 值太大，就会损坏器件。

**Substitution of Eq.(1－2) into Eq.(1－3)** yields Eq.(1－4).

若把式(1－2)代入式(1－3)，我们就得到了式(1－4)。

**A physical analysis of the diode** shows that….

若对二极管作一物理分析就能看到……。

**A little reflection** will show that….

如果稍加思考就能看出……。

**A comparison of water waves with radio waves** shows that⋯.

把水波与无线电波作一比较就可看出……。

**A thorough examination of these assumptions** reveals that⋯.

若对这些假设仔细考察一下，就能看出……。

【改正后的句子】　**A comparison with the MEI method shows that the third postulate in the MEI method is questionable[ is open to question].**

例 356

【汉语原句】　利用超分辨率方法来提高在 ISAR 成像中距离调整的精度

【英语错句】　Using super resolution technique to improve the accuracy of range alignment in ISAR imaging.

【错误分析】　(1)这是一篇论文的标题，根据英美人的写作习惯，最好是采用一个名词短语(尽管有个别人用一个动名词短语)，但不能含有任何从句，如"(A) study of the stability of an array cathode"(对于阵列阴极稳定性的研究)，也有少数人采用"on 短语"，意为"论……"，如 "On fault-tolerant routing in hypercubes(论超立方体网络中的容错路径选择)"。在本例中，应该把"to improve"改成名词短语"improvement of"，这是原意的侧重点，而把"using…"放在最后表示"improvement"的逻辑状语，在"using"前面也可加上介词"by"。(2)在"super resolution technique"前应加上定冠词"the"。

【改正后的句子】　**Improvement of the accuracy of range alignment in ISAR imaging by using the super resolution technique**

例 357

【汉语原句】　该系统的性能改善了 14～30 dB，前提是不增加该系统的复杂程度。

【英语错句】　The performance of the system has been improved 14～30 dB on conditions that the system had no increases in equipment complexity.

【错误分析】　(1)"条件是"的表达式是"on condition that⋯"，其中"condition"既不用复数形式，其前面也不用任何冠词。(2)"不增加该系统的复杂程度"完全写错了，在此不要根据汉语的字面意思来写，而应该根据其内涵来写，可写成"the complexity of the system remains unchanged"。(3)"14～30 dB"为名词短语作数量状语，在其前面最好加介词"by"(当然这个"by"是可以省去的)以便使读者一目了然。

【改正后的句子】　The performance of the system has been improved by 14～30 dB on condition that the complexity of the system remains unchanged.

例 358

【汉语原句】　测量内容可分为电容、电流和电压的测量。

【英语错句】　Measurements are generally divided into capacitance, current and voltage.

【错误分析】　这个句子属于中文式的英文，根据汉语原句，其意为"（这儿的）测量分三类：对电容的测量、对电流的测量和对电压的测量"。表示"分成三类"可使用"be of[fall into] three groups[classes; kinds; categories]"；"对……的测量"可表示成"the measurement (made) of…"。例如："电子测量有两种：对电子量的测量和对其他量的测量。"其英语表示法是"Electronic measurements are of two kinds: those made of electronic quantities and those made of other quantities."

【改正后的句子】　**Measurements are generally of three groups: those made of capacitance, those made of current and those made of voltage. [或: … those made of capacitance, current, and voltage, respectively.]**

例 359

【汉语原句】　本文将 RELAX 算法从数据域推广到相关域。

【英语错句】　This paper extends the RELAX algorithm from data domain to correlation domain.

【错误分析】　这一句的主要错误在于没有理解汉语的真实含义，不是论文"把……从 A 推广到 B"，而是论述了这一问题，所以应该把"extends"改写成"deals with the extension of … from A to B"。

【改正后的句子】　**This paper deals with the extension of the RELAX algorithm from the data domain to the correlation domain.**

例 360

【汉语原句】　从图 1 可以看出本方法的均方差（MSE）比 Park 的方法来得小，但是当 SNR 大于 15 dB 时几乎就一样了。

【英语错句】　From Fig. 1 it can be seen that the MSE of this method is smaller than the Park's method, but it is almost same when the SNR is greater than 15 dB.

【错误分析】 （1）在"人名所有格＋名词"前不得用冠词。（2）主语从句中比较对象不一致，应该改成"…than that by Park's method"。（3）在"but"后的内容属于并列的第二个主语从句，根据英美人的习惯，在"but"后应该加上一个从句引导词"that"。（4）根据句子的含义，"but"后的"it is"应该是"they are"。（5）在"same"之前根据习惯应该加有"the"。

【改正后的句子】 **From Fig. 1 it can be seen that the MSE by this method is smaller than that by Park's method，but that they are almost the same when the SNR is greater than 15 dB.**

例 361

【汉语原句】 这里是有待于解决的一个问题：在这样一种不确定的条件下，如何对这种广告建模呢？

【英语错句】 Here is a problem remained to be solved：in such uncertain condition，how to model this advertising?

【错误分析】 （1）"remain"是不及物动词，所以不能用其过去分词作定语，应该使用现在分词"remaining"，不过绝大多数英美人喜欢用一个定语从句来表示这一含义，即"which remains to be solved"。（2）由于"condition"是可数名词单数，因此在"such"后面应该加上"an"。（3）在此不能用"how to model this advertising?"，应该使用一个完整的问句，即"How can one[we] model this advertising?"。

【改正后的句子】 **Here is a problem which remains to be solved：How can one/we model this advertising in such an uncertain condition?**

例 362

【汉语原句】 在反变换之前，接收的信号的平均功率被放大到与 $y_k$ 的平均功率相同。

【英语错句】 Before the inverse transform，the average power of received signal is amplified to be the same as that of $y_k$.

【错误分析】 （1）句中不能用"to be …"来表示"amplified"的程度，应该写成"to the extent that it is …"。（2）在"received signal"前应该有冠词。

【改正后的句子】 **Before the inverse transform，the average power of the received signal is amplified to the extent that it is the same as that of $y_k$.**

**例 363**

【汉语原句】　假设滤波器的系数为 $h(u)$，其阶为 $N$，其频率响应为 $H(j\omega)$。

【英语错句】　Supposing the coefficient of filter is $h(u)$, its order is $N$ and its frequency response is $H(j\omega)$.

【错误分析】　（1）由于本文不成句，所以不能用"Supposing"而应该用 "Suppose"，作为一个祈使句。（2）在"filter"前应该加定冠词。（3）由于动词 "suppose"后跟三个宾语从句，每个宾语从句前均应有一个引导词"that"（根据语法规则，第一个"that"是可以省去的）。

【改正后的句子】　**Suppose（that）the coefficient of the filter is h(u)，that its order is N，and that its frequency response is H(jω).**

**例 364**

【汉语原句】　这个数值在观察中出现的次数列在第一列中。

【英语错句】　The times this value occurs in the observations is given in the first column.

【错误分析】　"次数"应该为"the number of times"，不少人往往把 "number of"丢掉了。在英语（特别是在科技英语）中，在某些词语后的定语从句可以省去引导定语从句的关系副词或"介词＋which"，这时也可用起关系副词作用的"万能关系副词 that"来引导。常见的这些词语有：

the time，moment，instant，day 等：后面省去了 when、at［on；in；…］which［that］

the reason：后面省去了 why、for which［that］

the way，manner：后面省去了 in which［that］

the direction：后面省去了 in which［that］

the distance：后面省去了 through which［that］

the amount：后面省去了 by which［that］

the number of times［units；days；…］：后面省去了 by which［that］
例如：

The product of the force and the **distance** <u>a body moves</u> is called work.
力与物体运动的距离之乘积被称为功。

The **number of times** <u>this particle vibrates per second</u> is the frequency.
<u>这个质点每秒钟振动的次数</u>就是频率。

The **amount** a solid material will expand when heated is measured by its coefficient of linear expansion.

固体物质受热时膨胀的量是用它的线膨胀系数来度量的。

The damped oscillation depends on the **time** the switch is closed.

阻尼振荡取决于开关闭合的时间。

This diagram shows the **way** the current changes with time.

这个图说明了电流随时间变化的情况[方式]。

The **reason** this is called current feedback is that the amount of current flowing through $R_s$ determines the percentage of feedback.

把这种情况称为电流反馈的理由在于流过 $R_s$ 的电流大小确定了反馈的百分数。

Reflections come back only from objects in the **direction** the antenna is pointed.

回波只能从处于天线所指的方向上的物体那儿反射回来。

The voltage gain is the **number of times** a stage, or a number of stages, amplifies the signal.

电压增益就是一级或数级放大器放大信号的倍数。

【改正后的句子】　**The number of times this value occurs in the observations is given in the first column.**


例 365

【汉语原句】　我们已定义了度量 $m$ 的质量单位。

【英语错句】　We have defined mass units to measure $m$.

【错误分析】　本句的毛病出在不定式作定语上，这一错误也是读者常犯的，主要是不了解不定式作定语时与被定词之间一定要存在某种逻辑关系，多数情况下应该可以把该不定式短语扩展成一个定语从句，所以说不定式作定语可看成是一个定语从句的缩略形式。下面介绍一下不定式作定语时与被定词之间存在的四种主要逻辑关系。

①"主谓关系"(可以是主动式，也可以是被动式)。也就是说，不定式所修饰的名词是不定式所表示的动作的逻辑主语。例如：

This is a special-purpose computer **to solve only one kind of problem.** ($\approx$ ⋯which can solve only one kind of problem.)

这是一台只能解一类问题的专用计算机。(注："solve"这个动作是"computer"发出的，这是主动的主谓关系。)

The television camera scans the subject **to be transmitted**.($\approx\cdots$which will be transmitted)

电视摄像机对<u>要发送的</u>对象进行扫描。（注："the subject"与"to be transmitted"之间存在被动的主谓关系。）

Gallileo Galilei（1564—1642）was the first scientist **to understand clearly the concept of acceleration**.（$\approx\cdots$who understood clearly the concept of acceleration.）

加利莱奥·加利莱(1564年至1642年)是<u>清晰地理解加速度这一概念的</u>第一位科学家。

② "动宾关系"。这时不定式只能是主动形式，其前面的被修饰词就是不定式所表示的动作的承受者(即"逻辑宾语")，而动作的发出者一般就是句子的主语或泛指任何人。例如：

"Weight"is a difficult thing **to define**.（$\approx\cdots$that one can define.）

"重量"是一个难以<u>定义的</u>概念。（"thing"是动作"define"的对象）

From the physical standpoint the important thing **to remember**（$\approx\cdots$that we should remember）is that the weight of a body is the resultant of an infinite number of infinitesimal forces.

从物理学观点来说，我们<u>应该记住的</u>重要事情是：物体的重量是无限个无穷小的力的合力。

In 1932，physicists engaged in studying the atom and its nucleus had only three particles **to consider**：the proton，the neutron，and the electron.

在1932年，从事研究原子及原子核的物理学家们只有三种微粒<u>要考虑</u>：质子、中子和电子。

③ 不定式短语末尾的介词与其介词宾语的关系。这一逻辑关系是许多中国读者感到最难理解的，这时的不定式为主动形式，动作的发出者也是句子的主语或泛指的"人们"，这种不定式可以扩展成"介词＋which"开头的定语从句。例如：

There are many types of tools **to choose from**.（$\approx\cdots$from which one can choose.）

有许多种类型的工具<u>可供选用</u>。

A transistor must have a load **to work into** in order to develop a useful output.

晶体管必须要有一个<u>推动的</u>负载，以便产生一个有用的输出。

在正式的科技文中，这类不定式往往是以"介词＋which"开头的，如：

A horizontal line can be taken as the axis **along which to represent** $n$. ($\approx\cdots$ along which one can represent $n$.)

我们可以把一根水平线取作为表示 $n$ 所沿顺的轴。

In this case, the best choice for the axis **around which to calculate torques** is the base of the ladder.

在这种情况下，对计算力矩所围绕的轴的最佳选择是梯子的底部。

We shall use such a field **on which to base our discussion of magnetic properties.**

我们将使用这种场来作为讨论磁性质的基础。

若是一个不定式复合结构作定语的话，则只能把介词放在不定式复合结构尾部而不能在开头用"介词＋which"的形式，如：

Usually we cannot see the earth's shadow, because there is nothing **for it to fall on.**

通常我们是看不见地球的影子的，因为没有供影子落下的地方。

④ 纯修饰关系。这时不定式一般为主动形式，它修饰的是一些抽象名词，如 ability, capability, power, aim, purpose, objective, use, tendency, chance, opportunity, time, attempt, way 等。例如：

There are several ways **to take an average.**

取平均值的方法有几种。

The machine must have the ability **to print out the results obtained.**

该机器必须能够打印出所获得的结果。

There is no need **to plot numerous points.**

没有必要画出许许多多的点来。

This body has no tendency **to move or rotate in any direction.**

这个物体没有朝任何方向运动或转动的趋势。

We call the time **to complete one cycle** the period of the sinusoid.

我们把完成一周的时间称为正弦波的周期。

Imaginary numbers came into the science through the attempt **to obtain a solution of the quadratic equation in all cases.**

科学上之所以引入虚数是由于试图要获得二次方程式在各种情况下的解。

In solid materials, only some of the outer electrons may have any freedom **to move.**

在固体物质中，只有一部分外层电子可能有运动的自由。

总之，除了第四种逻辑关系外，若不存在主谓关系或动宾关系，则就应使用"介词＋which＋to（do）"的动词不定式形式。

【改正后的句子】 **We have defined mass units with which to measure *m*.**

### 例 366

【汉语原句】 这一技术能使传统检测应用于系统性能的评估。

【英语错句】 This technology can apply the traditional detection to the evaluation of system performance.

【错误分析】 本句属于中文式的英语，"technology"是不会发出"apply"这一动作的，应该采用"make it possible to（do sth）"。

【改正后的句子】 **This technology makes it possible to apply the traditional detection to the evaluation of system performance.**

### 例 367

【汉语原句】 导出了一个计算电荷分布的简单公式，最后举了若干例子。

【英语错句】 A simple formula to calculate the charge distribution is derived and some examples are given in the end.

【错误分析】 （1）在公式、算法、语言等词后的定语多用"for"短语。（2）"in the end"必须改成"finally"，不过在此可以省去不译。（3）第二个分句最好用"with"结构来表示，这样句子显得紧凑了。

【改正后的句子】 **A simple formula for calculating the charge distribution is derived，with some examples given.**

### 例 368

【汉语原句】 这条曲线显示了该电路中的电流随外加电压的变化情况。

【英语错句】 The curve shows the situation of the current in the circuit varying with the applied voltage.

【错误分析】 （1）注意固定的词汇搭配模式"the variation of A with B"（意为"A 随 B 而变化［的情况］"）。（2）"the situation"是不需要加上的。

【改正后的句子】 **The curve shows the variation of（the）current in the circuit with（the）applied voltage.（更好的版本为：The curve shows how the current in the circuit varies with the applied voltage.）**

### 例 369

【汉语原句】 在这种情况下有多类电路可供我们选用。

【英语错句】　In this case there are many types of circuit for us to choose.

【错误分析】　本句的错误是原作者可能不了解"choose"与"choose from"的用法的区别：前者表示"选择"，其宾语被全部选中；而后者则表示"从中选择一部分"，从本句的含意可明显地看出我们在此应该使用"choose from"。

【改正后的句子】　**In this case there are many types of circuit for us to choose from.**


例 370

【汉语原句】　这是由于这两个元件的结构不同引起的。

【英语错句】　This is caused by that this two components are different in structure.

【错误分析】　(1)"that"是不能引导介词宾语从句的，这时应该在介词后加上"the fact"。(2)"这两个"应该写成"these two"而不是"this two"，这主要是受了汉语"这"的影响而造成的错误，有些读者常犯这一错误。

【改正后的句子】　**This is due to the difference in structure between these two components.**


例 371

【汉语原句】　该译码器的结构比较简单，对(2，1，6)卷积码译码时，译码速率可达 147 Kb/s。

【英语错句】　The construction of the decoder is simple. The decoding speed for (2，1，6) convolutional codes reaches 147 kb/s.

【错误分析】　从英语句子的语法本身来说上面的表达并没有错误，但从科技英语的句子结构来说就显得十分松散，应该把其两部分合并在一个句子中，后一部分可以用"with 短语"来表示，写成"with the decoding speed for (2，1，6) convolutional codes of up to 147 kb/s."

【改正后的句子】　**The decoder is simple in structure，with the decoding speed for (2，1，6) convolutional codes of up to 147 kb/s.**


例 372

【汉语原句】　这个值似乎不能满足该方程。

【英语错句】　It appears that this value does not satisfy the equation.

【错误分析】　本句不能说是错误的，但按英语习惯最好否定在 appear 上，这叫"否定的转移"。在科技文中主要有以下三种情况：

① 主句动词为 "think，feel，suppose，expect，believe，imagine，reckon" 等时，他们的否定形式实际上是否定了后面的东西。例如：

I **do not think** that this equation holds here.

我认为这个式子在此<u>不适用</u>。

② 主句动词为 "happen，appear，seem" 时，他们的否定形式实际上是否定了后面的东西。例如：

This transistor **does not seem** to be so good as that one.

这只晶体管似乎<u>不</u>如那只好。

③ 主句为否定式（含有 "not，no，rather than，instead of" 等）而从句为由 "as" 引出的方式状语从句、特殊短语或非限制性定语从句时，否定的含义处于 "as" 从句等上。例如：

This disease **does not affect** animals as it does humans.

这种病并<u>不像</u>它影响人类那样影响动物。

The Maxwell relations **are not modified** by relativity as Newton's relations are when particle velocities approach the speed of light.

当质点速度趋近光速时，麦克斯韦关系式并<u>不像</u>牛顿关系式那样要用相对论加以修正。

【改正后的句子】 **It does not appear that this value satisfies the equation.**

**例 373**

【汉语原句】 为了与所提出的方案进行比较，本文对其他三种 OFDM 系统进行了仿真。

【英语错句】 In order to compare with the proposed scheme, the other three OFDM systems are simulated in this paper.

【错误分析】 （1）"to compare" 没有它的宾语，所以应该改成 "to make a comparison"。（2）"other three" 应该改成 "three other"。

【改正后的句子】 **In order to make a comparison with the proposed scheme，the three other OFDM systems are simulated in this paper.**

**例 374**

【汉语原句】 本文主要论述对这种信号进行 Wiener 滤波以及用于它们的 LMS 算法.

【英语错句】 This paper mainly deals with the Wiener filtering and LMS algorithm of such signals.

【错误分析】 (1)本句主要错误在于"filtering"和"algorithm"后面应跟的介词是不同的，前者后面应跟"of"来表示"动宾关系"，而后者后面应跟"for"来表示适用的对象。(2)副词"mainly"应该放在"with"之前；(3)在"LMS"前应加一个定冠词。

【改正后的句子】 **This paper deals mainly with the Wiener filtering of and the LMS algorithm for such signals.**

### 例 375

【汉语原句】 特别在无线电传输领域，这设备有许多优点。

【英语错句】 Specially in the field of radio transmission, there are many advantages with the device.

【错误分析】 (1)"特别"在此处应该使用"especially"而不是"specially"，后者主要表示"专门"之意。(2)这里的"有"应该用动词"have"来表示。

【改正后的句子】 **Especially in the field of radio transmission, the device has many advantages.**

### 例 376

【汉语原句】 这样，运作开销和通讯开销都降低了，从而效率就提高了。

【英语错句】 In this way, operation and communication overheads are reduced, this make the schemes very efficient.

【错误分析】 为了表示"都"的含意，应该在"operation"前加上"both"。如果仅从语法上来说，在"this"前应该加上"and"，并且要把"make"变成单数第三人称形式，即要在末尾加"s"。不过为了使句子紧凑且精炼，应该把逗号后的部分译为分词短语作结果状语更好(当然也可以把"this"改成"which"也行，"make"同样要用"makes"，但这一句型不够精练)。

【改正后的句子】 **In this way, both operation and communication overheads are reduced, thus making the schemes very efficient (或: thus raising the efficiency).**

### 例 377

【汉语原句】 该模型成功地解释了基区表面电流随栅极电流变化的实验曲线。

【英语错句】 The model successfully explains the experiment curves of the base surface current varying with the gate current.

**【错误分析】**　（1）本句中的主要错误在于没有正确地表达"A 随 B 的变化"这一概念，应该使用固定的词汇搭配模式"the variation of A with B"。（2）"实验"在此应该用形容词"experimental"来表达。（3）"成功地……"可有三种表示方式："successfully do…""succeed in doing…""be successful in doing…"。

**【改正后的句子】**　**The model successfully explains the experimental curves of the variation of（the）base surface current with（the）gate current.（也可以表示为…curves of（the）base surface current against/versus（the）gate current）**

　　**例 378**

**【汉语原句】**　用线性比相方法进行器件及设备的时延测量，不但测量精度高，而且设备构成相当简单。

**【英语错句】**　Measuring the time delay of devices and equipments with the method of linear phase comparison, the measuring accuracy is high and the instrument is very simple.

**【错误分析】**　本句的主要错误在于作状语的分词短语中"measuring"的逻辑主语并不是句子的主语"the measuring accuracy"，这属于"垂悬分词"，其修改方法有两种：一是把分词短语改成条件状语从句"If one measures the time delay…"，同时最好在"accuracy"后加上过去分词"obtained"和在"instrument"后加上过去分词"used 或 required"；二是改变句子的主体部分（即分词短语后面的部分），写成"one can obtain a high accuracy（only）by using a very simple instrument"。

**【改正后的句子】**　**If one measures the time delay of devices and equipments by/with the method of linear phase comparison, not only is the measuring accuracy obtained high, but also the instrument used is rather simple.**

　　**例 379**

**【汉语原句】**　通过处理输入所获得的正认知效果越大，它就越相关。

**【英语错句】**　The more positive cognitive effects achieved by processing an input, the more relevant it will be.

**【错误分析】**　（1）与"effect"相搭配的形容词应该是"great"，所以这里应该把"more"改成"greater"。（2）在"achieved"前应该加上"are"，这样本句的第一部分才是正确的。（3）"它"在此实际上是指"它们"。

**【改正后的句子】**　**The greater positive cognitive effects are achieved by processing an input, the more relevant they will be.**

**例 380**

【汉语原句】 通过该研究可以得出结论：在这种情况下，可靠性问题主要是由电子通过氧化层隧穿引起的。

【英语错句】 Conclusion can be drawn from this study: in this case the reliability issue is mainly caused by electrons tunneling via oxide.

【错误分析】 (1) 在"conclusion"之前应该有定冠词。(2) 汉语句中的冒号可以用"that"来代替，它引导一个同位语从句。(3) "electrons tunneling via oxide"用"the tunneling of electrons through the oxide"代替为好。

【改正后的句子】 **The conclusion can be drawn from this study that in this case the issue of reliability is mainly caused by the tunneling of electrons through the oxide.**

**例 381**

【汉语原句】 加权顺序统计滤波器的新的表示法及优化

【英语错句】 A new expression and optimization for weighted order statistic filters

【错误分析】 这是一篇论文的标题，错误在于两个名词"expression"和"optimization"之后应该使用不同的介词，前者一般跟"for"而后者要用"of"来引出其逻辑宾语。

【改正后的句子】 **A new expression for and the optimization of weighted order statistic filters**

**例 382**

【汉语原句】 伽马射线的本质与阿尔法射线是完全不同的。

【英语错句】 The nature of the gamma rays is quite different from the alpha rays.

【错误分析】 (1) 在两种射线前面的定冠词要去掉。(2) 本句的比较对象不一致，在"alpha rays"前应该加上"that of"。不过本句最好写成英美科技人员常用的句型"A is quite different in … from B"。

【改正后的句子】 **Gamma rays are quite different in nature/essence from alpha rays.**

**例 383**

【汉语原句】 利用 8 点法，振动的范围可能约高达一万像素，所以该方法有时是没有用的。

【英语错句】　By the use of the 8-point method，the range of vibration may be about ten thousand pixels. So it is useless sometimes.

【错误分析】　（1）对于"范围"的"高达"应该使用"as wide as…"来表示。（2）为了使句子紧凑，应该使用"so that"从句把英语的两个句子合并成一个。（3）句尾的"sometimes"应该改为"in some cases"为好。

【改正后的句子】　**By the use of the 8-point method the range of vibration may be as wide as about ten thousand pixels so that the method is useless in some cases.**

例 384

【汉语原句】　本文主要论述有关空气污染物的信息和对它们的处理。

【英语错句】　This paper mainly deals with the information and treatment of the air pollutants.

【错误分析】　（1）根据英美人的习惯，应该把副词"mainly"放在不及物动词和与它搭配的介词之间。（2）在"information"后应该跟介词"on"或"about"而不能跟介词"of"的。（3）由于"空气污染物"在此并不是特指，所以在其前面不用定冠词。

【改正后的句子】　**This paper deals mainly with the information on and the treatment of air pollutants.**

例 385

【汉语原句】　这个算法能够计算出化学剂对它们所处环境起作用的能力。

【英语错句】　This algorithm can calculate the ability of agents in acting on the environment that they exist.

【错误分析】　（1）"can calculate"应改成"can be used to calculate"。（2）"in acting"应该改成"to act"，这属于固定表示法"the ability of A to do B"。（3）"that"在此不能引导定语从句，应该把它改成"where 或 in which"。

【改正后的句子】　**This algorithm can be used to calculate the ability of agents to act on the environment in which they exist.**

例 386

【汉语原句】　医院管理信息系统查询优化的研究和实现

【英语错句】　The research and realization of the enquiry and optimization of hospital management information system

　　【错误分析】　这是一篇论文的标题。(1)"research"后面是不能跟"of"的，一般是跟"on"。(2)"查询"是"优化"的对象，而且实际上在此"查询"是"查询功能"的意思。(3)在"……系统"前应该有冠词。

　　【改正后的句子】　**Research on and realization of（the）optimization of the enquiry function of a hospital management information system**

　　例 387

　　【汉语原句】　该系统的特点是体积小、操作简单、成本低。

　　【英语错句】　The features of this system is small in size, easy in operation, low in cost.

　　【错误分析】　(1)由于"特点"作主语而不是某个东西作主语，所以不能用"形容词＋in …"，而只能用名词短语表示，这是有些读者常犯的错误。(2)"is"应该改为"are"。(3)在"low"之前应该加"and"。

　　【改正后的句子】　**The features of this system are small size, easy operation and low cost.**（更好的版本为：**This system is characterized by（its）small size, easy operation and low cost.**）

　　例 388

　　【汉语原句】　对于复杂的视频输入，观察到速度提升最高可达 100 倍以上。

　　【英语错句】　The observed speedup reaches more than 100 X for complex video inputs.

　　【错误分析】　(1)"观察到"应该写成"It is［has been］observed that"。(2)后面部分应该写成"the highest speed can be increased by 100 times"。

　　【改正后的句子】　**It is observed that the highest speed can be increased by 100 times for complex video inputs.**

　　例 389

　　【汉语原句】　这样就可以完成对原信号的识别任务。

　　【英语错句】　So identification tasks of the original signal could be solved.

　　【错误分析】　(1)"这样"一般译成"in this way"。(2)"对原信号的识别任务"应该译成"the task of identification of the original signal"，"任务"应该用单数形式。(3)"可以完成"一般译成"could［can］be accomplished［done］"。

【改正后的句子】 **In this way, the (task of) identification of the original signal could/can be accomplished.**

例 390

【汉语原句】 计算距离谱的这种方法可用来进行优化设计或理论分析。

【英语错句】 This method of computing distance spectrum may be used to design optionally or analyze theoretically.

【错误分析】 本句也属于中文式的英语句, 在及物动词"design"和"analyze"后没有宾语。(1) 在"method"后最好用"for"。(2) 在"distance spectrum"前加上"the"。(3) "优化"应该是"optimization"。(4) "to"后面应改为"for optimization design or theoretical analysis."。

【改正后的句子】 **This method for computing the distance spectrum may be used for optimization design or theoretical analysis.**

# 附　　言

## Ⅰ、冠词

在写作时，冠词是经常写错的，主要是三方面的问题：（1）丢失冠词。这是最大量的错误，基本上在每篇文章（甚至在每篇文摘）中均有不少这方面的问题。（2）在某些特殊情况下不知道冠词要用特殊位置。（3）不清楚用"a"和"an"的场合。

冠词的用法是比较复杂的，但下面列出的一些基本知识是不难掌握的。

### 1. 应该使用冠词的最基本的规则

（1）在普通的单数可数名词前一定要有冠词。泛指时多用不定冠词；特指的名词前要用定冠词。例如：

晶体管是由三部分构成的。

**A** transistor consists of three parts.

至于到底用"an"还是"a"，完全取决于不定冠词后面那个词的第一个音素（而不是字母）是元音还是辅音。如果是元音则要用"an"，否则用"a"。（它与字母无关，这一点也是不少读者在写论文时经常搞错的。）例如：

Here is **a** simple experiment.

下面是一个简单的实验。

（或：下面我们来做一个简单的实验。）

This is **a** unit for measuring weight.

这是一个（用于）测量重量的单位。

（注意字母"u"的第一个音是辅音[j]，所以其前面用冠词"a"。）

A magnet has **an** S pole and **an** N pole.

磁铁有一个 S 极和一个 N 极。

（注意字母"s"和"n"的第一个音是元音[e]，所以其前面要用"an"。）

If **a** voltage is applied across **the** terminals of **a** closed circuit，**a** current will flow in **the** circuit.

　　如果把(一个)电压加在(一个)闭合电路的两端,在(该)电路中就会有电流流动。

　　(在"terminals"前用了"the"是因为其后有一个后置定语"of a closed circuit",所以这个"terminals"是特指;在第二个"circuit"前加了"the"是指前面已提到的"a closed circuit",所以属于"特指"。)

　　但在不少情况下,英美科技人员在特指的复数名词前可以不用定冠词。例如:

It is necessary to measure the potential difference across two terminals of the battery.

　　(我们)必须测出该电池两端的电位差。

　　当然,怎么判断"强调特指"与否,完全取决于写作者自己的意愿,也就是说,有时加不加定冠词一般都是可以的。下面举一本美国中学化学书上在同一段中类似的两个句子,一个加了定冠词,而一个则没有加定冠词:

**The elements** in Group 1 are called alkali metals.

第一组中的元素被称为碱金属。

**Elements** in the same main group show very similar chemical reactions.

同一大组中的元素显示出非常类似的化学反应。

　　在带有后置定语而不强调特指的可数名词单数前,也可以使用不定冠词表示一类。例如:

化合物是能够分解成两种或多种元素的(一种)纯净物质。

A compound is **a** pure substance that can be broken down into two or more elements.

　　代词"one"后有定语从句时,它前面有时加定冠词、有时则不加。例如:

An a.c. is **one** whose direction is reversed at regular intervals.

交流电是其方向以规则间隔反向的电流。

The most widely encountered optical beam in quantum electronics is **the one** where the intensity distribution at planes normal to the propagation direction is Gaussian.

　　在量子电子学中最常遇到的光束,是垂直于传播方向平面处的强度分布是高斯型的那种光束。

　　(2) 对于没有被限定的物质名词或抽象名词,一般不加冠词;如果是表示一个被定义(或被称呼)的术语而且是单个抽象名词的话,一般不用定冠词(但也有少数人用定冠词的);如果是一个名词短语(单数形式),则一般也要用定冠词。例如:

Information is very important in the modern world.

在现代世界上,信息是非常重要的。

The rate at which work is done is called power.

做功的速率被称为功率。

This quantity is called (the) diffusivity.

这个量被称为扩散率。

The acceleration of a freely falling body near the earth's surface is referred to as the acceleration of gravity.

自由落体在地球表面附近的加速度被叫做重力加速度。

注意 1:不少读者一见到汉语句子中有词语"一种"时,马上就写成"a kind (sort;type)of …",往往是错误的。根据我的体会,其一般的规律是:如果有一系列的多个东西,特别是一种产品,那么可以用"a kind(sort;type) of";如果只有一个东西时,如"方法"、"语言"、"算法"、"方案"、"技术"等等,一般就不能用"a kind of"。例如:

This is a new kind of engine.

这是一种新的发动机。

This is a new algrithm.

这是一种新的算法。

注意 2:表示"一种(套)新的……"时,形容词"新的(new);特殊的(special)"等在英语中只能放在"kind;type;sort;set"等词之前,而不能放在"of"后的名词前("of"后不能有冠词,且该名词一般用单数),这也是读者经常犯的一个错误。而且该名词一般用不带冠词的单数形式(也有人用复数形式的),如:"一种新的仪器"应该写成"a new kind of instrument";"一类特殊的机器人"应该写成"a special type of robot";"一套旧的工具"应该写成"an old set of tool"。

(3) 对于心目中特指的东西、带有后置修饰语(包括单词、短语或从句)的东西、前面已提到过的、同一句中第二次提到的东西、或世上独一无二的东西(如"太阳""地球""月亮""宇宙"等)前应该使用定冠词。例如:

The voltage between the base and the emitter is small.

基极与发射极之间的电压是很小的。

**The** earth is one of the planets.

地球是行星之一。

## 2. 科技写作中不加冠词(或省去冠词)的常见情况

(1) 除了上面提到的泛指的物质名词及抽象名词前不加冠词外，表示一类的泛指的复数名词前不加冠词。例如：

**Satellites** play a very important role in modern communications.

卫星在现代通信中起着十分重要的作用。

(2) 论文的标题、书籍的各节的名称等开头的冠词可以省去[现在国内学报要求论文标题第一个名词前的冠词要省去]。例如：

5.1　Eye diagram

第5.1节　眼图(观察图)

6.3　Indirect method

第6.3节　间接法

34－6　(The)**Quantum theory** of light

第34－6节　光的量子理论

但标题若是可数名词的话，多数用复数形式表示，也有用定冠词加单数名词表示。例如：

5－2　Electron **orbits**

第5－2节　电子轨道

3－11　**The** telescope

第3－11节　望远镜

(3) 专有名词前的情况。在科技文写作中，专有名词主要涉及人名、地名、单位或机构名称和国家名称等。下面分别加以说明。

① 在人名前不加冠词，如：Faraday(法拉第)，Einstein(爱因斯坦)，Mao Zedong(毛泽东)，Xi Jinping(习近平)(请注意我国人名的拼写方法，这是我国政府规定的，发现读者在论文中写法比较乱。)

② 在人名带有头衔或职称时，在头衔和职称前不得加冠词，且头衔或职称的第一个字母要大写(这也是读者写作时偶尔会出错的)。例如：Chairman Xi(习主席)，**Professor** Wang(王教授)。

③ 地点名词前一般不加冠词，如：London(伦敦)，New York(纽约)，Beijing(北京)，Xi'an(西安)(注意：由于拼音的关系，在"i"和"an"之间要加

"'"，表示这两者不能拼读在一起；同样，"延安"应该写成"Yan'an"），"Shaanxi（陕西）"（注：由于英语中没有四声，所以无法区分"陕西"和"山西"，我国特规定，"陕西"的英语为"Shaanxi"，即比"山西""Shanxi"的拼写多了一个字母"a"以示区别）。

④ 国家名称或单位、机构名称需要注意以下两点：

ⅰ. 只由一个单词表示的国家名称绝大多数不加定冠词，但由多个词构成的国家名称或单位、机构名称前要加定冠词（不过写在信封上或在发表的论文作者下面等场合时一般可省去冠词）。例如：

China
中国
**the** People's Republic of China
中华人民共和国
America
美国
**the** United States of America
美利坚合众国
**the** Chinese Academy of Sciences
中国科学院
**the** School of Computer Science
计算机科学学院
**the** Westinghouse Electric Company/Corporation
西屋电气公司

ⅱ. 有关大学名称的情况：凡是"大学（university）"单词前只有专有名词的，其前面不得加冠词。例如：

Tsinghua University
清华大学
Xidian University
西电大学（西安电子科技大学的英文名称）
Xi'an Jiaotong University
西安交通大学
Peking University
北京大学

（注：由于"北京大学"的英语名称"Peking University"早就为世人所熟知了，所以不写成"Beijing University"；又如"香港"为"Hong Kong"，"澳门"为"Macao"。）

如果某大学名称是含有两个或两个以上普通词汇（"university"一词包括在内）时则要加定冠词。例如：

**the** University of Pennsylvania

宾夕法尼亚大学（宾州大学）

**the** Massachusetts Institute of Technology

麻省理工学院

**the** Northwestern Polytechnical University

西北工业大学

但个别也有例外，主要看外国人自己的习惯用法了。例如：

**the** George Washington University

乔治·华盛顿大学

（4）图示中一般可省去冠词（当然也可以加上冠词）。例如：

Fig. 1.1　**Keyboard** with **attached printer** and **oscilloscope display.**

图 1.1　键盘及其附带的打印机和示波显示器

（注：在"Keyboard"前面省去了"The 或 A"；在"attached printer"前面省去了"an"；在"oscilloscope display"前省去了"an"。）

（注：图示的说明不论是否是一个句子，末尾均要加上一个句号，即一个黑点。）

（5）某些可数名词单数形式泛指时可省去冠词（特别是在"between A and B"，"from A to B"，"the variation of A with B"，"the response of A to B"等等表达式中，在"A"和"B"之前可以省去定冠词）。例如：

This characteristic is not uniform from **diode** to **diode.**

各个二极管的这种特性并不是均匀的。

The variation of **output** with **input** is shown in Fig.1.

输出随输入的变化情况示于图 1 中。

A transistor consists of three parts：**emitter**，**base** and **collector.**

晶体管是由发射极、基极和集电极三部分构成的。

**Experiment** indicates that Ohm's law holds only for metallic conductors.

实验表明欧姆定律只适用于金属导体。

（但也有人写成：Experiments indicate that …；注意：句中"only"一般要放在不及物动词与介词之间，见"副词"部分。）

This problem can be solved by **computer**.

这个题可以用计算机来解。

（6）表示独一无二的人之前一般不用冠词（省去了一个定冠词）。例如：

This is called a hertz in honor of Heinrich Hertz, **discoverer** of radio waves.

这被称为 1 赫兹，以纪念无线电波的发现者海因里希·赫兹。

The unit of power is a joule per second, which is called a watt（W）, in honor of James Watt, **developer** of the steam engine.

功率的单位是 1 焦耳每秒，这被称为 1 瓦特（W），以纪念蒸汽机的研发者詹姆斯·瓦特。

如果不是"独一无二"的话，则通常要加上冠词。例如：

"Scientists see this as the last industrial moment," said Frank Y. Fradin, **a** physicist at the Argonne National Laboratory near Chicago.

"科学家们视之为最后的工业契机，"弗兰克·Y·弗拉丁说，他是位于芝加哥附近的阿贡国家实验室的物理学家。

（7）在人名的所有格之前不用冠词（至于何时用所有格、何时用普通格，完全要遵从英美人的习惯，一定要在阅读时善于观察）。例如：

**Newton's** laws of motion.

牛顿运动定律

**Ohm's** law.

欧姆定律

但如果人名直接修饰普通名词时，一般在它之前要用定冠词，例如：

**the Laplace** equation

拉普拉斯方程

**the Joule** effect

焦耳效应

（8）方程、表达式、公式、图表、章节、页码等等后跟数字表示"第……"时，其前面不得加冠词，而且第一个字母要大写（page 除外，"p"要小写）。例如：

**Equation（2 - 1）**

方程(2-1)

**Chapter 1**

第一章

**Section 3.1**

3.1 节

**Table 2-5**

表 2-5

**Fig. 1.2**

图 1-2

（注意：在图示时，在图的标号后不得加一点，而是空一格接着写图示的内容。这也是不少读者写作时往往犯的一个小错误。）

**page 5**

"第 5 页"

（9）在解释方程、公式、表达式等里面的参数时，一般可以不用冠词。例如：

$s = vt$

where $s = $ **distance**

$v = $ **velocity**

$t = $ **time**

$s = vt$，式中 $s$ 表示距离，$v$ 表示速度，$t$ 表示时间。

（注意：这里的关系副词"where"要顶格写，不能退后几格后开始写，因为这是属于上下主句和从句的关系，是一个复合句，"where"是引导修饰上面表达式或公式的非限制性定语从句，所以其第一个字母"w"要小写。这也是不少读者写论文时经常写错的。）

（10）在"by A is meant B"的定义句型中，"A"之前多数人不用冠词。例如：

By **frequency** is meant the number of times something repeats itself per second.

频率指的是某事每秒钟自己重复的次数。

By **linear operation** is meant the ability of an amplifier to amplify signals with little or no distortion.

所谓线性工作指的是放大器以很小的失真或无失真地放大信号的能力。

（11）在学科名称（mathematics，physics，electronics 等等）前不用冠词。例如：

**Mathematics** is a very useful tool in science.

数学是科学上非常有用的工具。

（注意：当遇到"the mathematics"时，则一般表示"数学内容/知识"）。

（12）表示"在某一方面"时一般不用冠词。例如：

Computers differ greatly **in size**.

各计算机的体积差别很大。

This machine is very good **in performance**.

这台机器性能很好。

希望读者要学会这一表示法。我们一般都习惯于写成"the sizes of computers"，"the performance of this machine"等等，尽管语法正确，但不属于英美科技人员偏爱的习惯表达法。

另外，英美科技人员用"in … form"和"in … order"，其中不用定冠词。但是，在用"domain"时，却要在其前面用定冠词；在用"fashion"时可用不定冠词。例如：

The number is transmitted to the receiver **in binary form**.

该数以二进制形式传送到接收机。

This integral can be evaluated **in closed form**.

这个积分可以按闭合形式求出来。

They are arranged **in alphabetic order**.

它们按字母表顺序排列。

After transformation，the signal **in the time domain** is changed into the signal **in the frequency domain**.

经过转换，时域信号转变成了频域信号。

Doing so will modify the signal **in a deterministic fashion**.

这样处理会以确定的方式修正信号。

（13）在"per（每）"之后不用冠词。例如：

Radio waves travel 300 000 kilometers per **second**.

无线电波每秒钟传播 30 万公里。

## 3. 可用可不用的情况

在用"系表"结构定义某个参数或名称时，可以不用冠词，也可以用冠词。例如：

**Absolute error** is the actual difference between the measured value and the accepted value.

绝对误差是测得值与认可值之间的实际差别。

**A scalar quantity** is one that possesses magnitude only.

标量是只具有大小的一个量。

**Pollen tube** is defined as the tube through which pollen passes into the ovary.

花粉管被定义为花粉进入子房的管道。

## 4. 一般要用定冠词的特殊情况

（1）表示某个参数的单位时，一般要用定冠词。例如：

The unit of resistance is **the ohm**.

电阻的单位为欧姆。

The unit of capacitance is **the farad**.

电容的单位为法拉。

（2）带有同位语的参数等名词前多数使用定冠词。例如：

系数 $\mu$ → the coefficient $\mu$

变量 s → the variable s

下标 $i$ → the subscript $i$

营销工具 $p$ → the marketing tool $p$

（3）在"same"前习惯上都要用定冠词。例如：

In this case, its potential difference is **the same** as the emf of the battery.

在这种情况下，其电位差就与电源电动势相同。

These two names are **the same** in meaning.

这两个名称的含义是相同的。

（4）在以下情况一定要用定冠词（不少读者往往出错）。

$$
\left.\begin{array}{l}
\text{Any} \\
\text{None} \\
\text{Neither} \\
\text{Either} \\
\text{All} \\
\text{Most} \\
\text{One} \\
\text{Each} \\
\text{The rest}
\end{array}\right\} \text{of } \textbf{the} + \text{名词（可数名词复数）}
$$

例如：

All of **the** devices here are home made.

这里的所有设备都是国产的。

None of **the** texts available mention this problem.

现有的教科书均没有提到这个问题。

If neither of **the** foregoing conditions is satisfied，then Eq.（4 – 1）is not readily solved.

如果前面两个条件都不满足，则式（4 – 1）不容易解。

**注意：**我们也见到下面的情况。

**The** $\varphi_i(t)$ form an orthonormal set.

"这些 $\varphi_i(t)$ 值构成了标准正交集。"

## 5. 几个名词并列时可以共用第一个名词前的冠词

例如：

Here D1 is replaced by a **silicon and germanium diode** in series.

这里 D1 由串联在一起的一只硅二极管和一只锗二极管所替代。

After the **hot and cold water** have been mixed，we have 2 kg of water at a temperature of 50℃.

在把冷水和热水混合后，我们得到了 2 千克温度为 50℃的水。

These convenient electric signal sources are very useful to the **test，maintenance，and operation** of various electrical appliances.

这些方便的电信号源，对各种电气设备的测试、维护或工作是很有用的。

## 6. 在许多不可数的抽象名词前使用不定冠词的情况

（1）在表示"学习一下""比较一下""计算一下""了解一下""考察一下""作一比较""作一分析""作一研究""作一讨论"等时，用不定冠词。例如：

We shall begin with a **study** of forces.

我们先研究一下各种力。

A **comparison** of Fig. 5.7 with Fig. 5.1 reveals that the binary and M-ray schemes are very similar.

把图 5.7 与图 5.1 作一比较，揭示出二进制方案和多元方案是非常相似的。

Let us make a **rough estimate** of the pressure there.

让我们粗略地估算一下那儿的压力。

A **brief qualitative discussion** of some basic concepts is presented in this paper.

本文简要定性地讨论了一些基本概念。

A **quantitative analysis** of this circuit is rather involved.

对该电路作一定量的分析是相当复杂的。

A **general knowledge** of the characteristics of electrical transmission is essential if the reader is to gain an **understanding** of data communications.

若读者想要懂得数据通信，就必须对电传输的特性具有一般的知识/了解。

（注意：有时候即使在同一本书中，同一作者在不同地方也有不用冠词的情况。）例如：

When coherent detection is used, **knowledge** of both the frequency and phase of the carrier is necessary.

当采用相干检波时，必须要了解载波的频率和相位。

**Inspection** of Fig. 3 − 8 shows that the two signals differ in phase.

审视一下图 3 − 8 可看出，两个信号的相位是不同的。

（2）在名词"increase, decrease, rise, fall, reduction, drop"等等前往往用不定冠词，而且它们后面往往跟"in"而少用"of"。例如：

An **increase** in pressure always causes a **decrease** in volume.

压力的增加总会引起体积的下降/缩小。

The Early effect causes **a 2% decrease** in $q_B$, and the Kirk effect causes a **4% increase**, producing a **net increase** of 2%.

厄雷效应使 $q_B$ 降低 2%，而柯克效应使它提高 45，从而使它纯增加 2%。

（3）少数固定句型中要用不定冠词。例如：

There is a **growing**（an **increasing**）**awareness** that it is urgently necessary to reduce global warming.

现在人们越来越认识到迫切需要降低全球性变暖。

## 7. 序词使用冠词的情况

在序数词并不强调次序、往往表示"另一个"的含义时，不用定冠词而用不定冠词，但仍可汉译成"第一""第二""第三"等。例如：

An important objective of **a first** course in mechanics is to train the student to think about physical phenomena in mathematical terms.

力学中第一门课程的一个重要目的，在于训练学生来思考用数学术语表示的物理现象。

As **a first** step in deriving the optimum filters, let us derive an expression for the probability of a bit error.

作为设计最佳滤波器的第一步，让我们推导出概率为一个比特误差的表达式。

**A second** approach is as follows.

第二种/另一种方法如下。

The advent of electronics is reckoned from the discovery that the current in a vacuum diode can be controlled by introducing a third electrode.

电子学的诞生是从发现了真空二极管中的电流可以通过引入第三个/另一个电极来加以控制算起的。

**A third** distortion is called nonlinear distortion.

另一种/第三种失真被称为非线性失真。

## 8. 冠词的特殊位置

（1）定冠词的特殊位置。

$$\left.\begin{array}{l} \text{all} \\ \text{both} \end{array}\right\} + \textbf{the} + 复数名词$$

例如：

All **the** aircraft here are home-made.

这里所有的飞机都是国产的。

（注意"aircraft"一词的单、复数同形。）

Both **the** terms are positive.

这两项均为正。

（2）不定冠词的特殊位置。

这是许多读者写作时很不熟悉的内容之一，主要应注意以下这一点：

$$\left.\begin{array}{l} \text{too（太）} \\ \text{so（如此）} \\ \text{as（如此）…（as）} \\ \text{how（多么）} \end{array}\right\} + 形容词 + a(an) + 单数名词$$

例如：

Too great **a** value of △ precludes self-oscillation.

△ 值太大能排除自激振荡。

It is necessary to determine how large **a** force is required to lift this body.

必须确定为提起这个物体需要多大的力。

（注意：在物理学中，"force"这个名词一般是可数名词。当把它看作一个抽象概念时，可作为不可数名词，这时可用"how much force"。）

This manipulator can lift as heavy **a** weight as 450 kilograms.

这只机械手能提起重达 450 公斤的重物。

## Ⅱ、词的搭配

词的搭配是英语学习中一个极为重要的项目，它对科技英语写作是特别重要的。各种搭配关系要靠读者在长期学习和阅读过程中逐个记忆、不断积累才能掌握好。在此只能纲要性地提出以下几点，供读者参考。

通常有以下两种搭配关系。

1. 固定词组

这些固定词组的数量极多，在大、中型词典中均用黑体或斜体标出，它们的构成是不得变动的，词义是固定的。例如：

| | |
|---|---|
| in other words | 换句话说 |
| take place | 发生；举行 |
| take the place of | 代替 |
| for example | 例如 |
| bring to light | 揭露 |
| in（the）light of | 由于 |

2. 习惯性的搭配

这种搭配尚未成为固定词组，要在文章、词典提供的例句中去体会和寻找。

（1）某些名词前要与特定的动词相搭配。例如：

| | |
|---|---|
| improve the quality | 提高质量 |
| raise the efficiency | 提高效率 |
| increase the ability | 提高能力 |
| enhance the vigilance | 提高警惕 |

从上面的例子可以看出，汉语中同一个"提高"，而在英语中要根据不同的名词用不同的动词搭配。

（2）某些动词、形容词、名词后要跟有特定的介词。

① 动词后。例如：

Laser beams **are emitted at** the moon.

把激光射束射向月球。（这里的"at"与"emit"连用时表示"朝向"之意。不少读者往往把它写成"to"或"toward"，这是错误的。）

类同的情况还有：

| throw … at | 把……扔向…… |
| aim … at | 把……瞄准…… |
| shoot … at | 把……瞄准……（射击） |
| send … at | 把……发向…… |
| point … at | 把……指向…… |

② 形容词后。例如：

This model **is descriptive of** the behavior of an atom.

这个模型描述了原子的情况。

类似的情况有：

| (be) representative of＝represent | 表示；代表 |
| (be) characteristic of＝characterize | 表示的特征 |
| (be) indicative of＝indicate | 表明 |
| (be) symptomatic of＝symptomize＝symptomatize | 表明；是……的症状 |
| (be) aware of＝know | 知道，了解 |

又如：

| (be) similar to | 类似于 |
| (be) equal to | 等于 |
| (be) equivalent to | 等效于 |
| (be) perpendicular to | 垂直于 |
| (be) vertical to | 垂直于 |
| (be) normal to | 垂直于 |
| (be) parallel to | 平行于 |
| (be) proportional to | 正比于 |
| (be) tangent/tangential to | 正切于 |

③ 名词后。例如：

That professor will deliver a **lecture on** mobile communication.

那位教授将作有关移动通讯方面的讲座。

She gives undergraduates **lessons in** the C＋＋ programming language.

她给本科生上 C＋＋程序设计语言课。

These students are making an **experiment in** physics.

这些学生在做物理实验。

（"in"后一般跟大的学科）

They are doing an **experiment on** electricity.

他们在做电学实验。

（"on"后跟比较小的学科，"电学"属于"物理学"的一部分。在"expert"后介词跟法与"experiment"类同。）

The professor has played an important role in the **research on** the Walsh Functions.

该教授对沃氏函数的研究起了重要的作用。

（在"research"后要跟"on"，也可跟"into，with"等，但不能跟"of"。）

④ 某些名词前要用特定的介词。例如：

Radio waves travel **in all directions.**

无线电波朝四面八方传播。

（表示"朝……方向"就要用"in"。）

**For this purpose**，we must connect a capacitor across the load.

为此，我们必须在负载两端接一个电容器。

（在"purpose"前用"for"。）

**At this temperature** the metal will melt.

在这个温度上该金属就会熔化。

（在"temperature"前用"at"。）

The expert will read an important academic paper **at the conference.**

该专家将在会议上宣读一篇重要的学术论文。

（在各种大小会议上均要用"at …"。）

⑤ 名词与介词的一个重要搭配模式。

有些读者对这模式并不熟悉。若能熟练地掌握这种搭配模式，就能使英汉互译通顺而地道。这种搭配关系的模式是：

名词＋of＋A＋介词＋B（这里的 A 和 B 也是名词）

对于这种搭配关系，下面分三种情况来说明，望注意每种情况的汉译法。

ⅰ. 上述模式中的那个"名词"是一般的抽象名词（或与形容词同根的抽象名词）。整个模式一般汉译成"A ……B 的……"。例如：

Speed is the **ratio** of distance **to** time.

速度等于距离与时间之比。

The **distance** of the sun **from** the earth is very great.

太阳离地球的距离是很远的。

The **effect** of temperature **on** conductivity of metals is small.

温度对金属导电率的影响是很小的。

This section deals with the **advantages** of transistors **over** electron tubes.

本节论述晶体管与电子管相比的优点。

"log $n$" denotes the **logarithm** of $n$ **to** the base $e$.

"log $n$" 指的是 $n$ 以 $e$ 为底的对数。

The **absence** of an electron **from** a silicon atom can be thought of as a hole.

硅原子中缺少一个电子，可以看成是一个空穴。

The **superiority** of radar **to** ordinary vision lies in the greater distances at which seeing is possible with radar.

雷达优于普通视力，主要在于利用雷达所能看清的距离比较远。

ⅱ. 该"名词"来自于不及物动词，与其搭配的介词一般与原来动词所要求的介词相同。整个模式一般汉译成"A ……B 的……"。例如：

The **variation** of $g$ **with** latitude is due in part to the earth's rotation.

重力加速度 $g$ 随纬度的变化，部分原因是由于地球的自转。

（动词时为"vary with …""随……而变化。"）

The **dependence** of $y$ **on** $x$ is expressed by $y = f(x)$.

$y$ 对于 $x$ 的依从关系用 $y = f(x)$ 来表示。

（动词时为"depend on …""取决于……"。）

Ellipses are used to describe the **motions** of the planets **around** the sun.

人们用椭圆来描述行星绕太阳的运行情况。

（动词时为"move around …""绕……运动"。）

We have discussed the **passage** of an electric current **through** liquid solutions of acids, bases, and salts.

我们讨论了电流通过酸、碱、盐溶液的情况。

（动词时为"pass through …""通过……"。）

The **response** of a body **to** a net force $F$ is an acceleration $a$ proportional to $F$.

物体对净力 $F$ 的响应就是正比于 $F$ 的加速度 $a$。

动词时为"respond to …""响应于……"。

在由表示变化的不及物动词变来的名词（如：increase, decrease, rise,

fall，reduction，variation 等)的情况下，其与介词的搭配关系也可变成如下的形式：

$$名词＋in＋A＋with＋B$$

也就是说，把前面模式中的 of 换成了 in，当然用 of 仍然是可以的。例如：

In the troposphere there is a steady **fall** in temperature **with** increasing altitude.

在对流层，温度随着高度的增加而不断地下降。

A gradual **increase** in resistance **with** speed is characteristic of friction between the boat's bottom and the water.

船底与水之间的摩擦力的特点是，阻力随速度的增加而不断增大。

ⅲ. 该"名词"由及物动词变来，与其搭配的介词一般与动词时所要求的相同。汉译时，一般要把该名词译成及物动词。例如：

A **comparison** of Eq. (4) **with** Eq. (6) leads to the following relations.

把式(4)与式(6)相比较，我们就得到以下几个关系式。

(动词时为"compare A with B"，"把 A 与 B 相比较"。)

The **resolution** of a force **into** $x$- and $y$-components is possible.

我们可以把一个力分解成 $x$ 分量和 $y$ 分量。

(动词时为"resolve A into B"，"把 A 分解成 B"。)

**Exposure** of the body **to** potentially toxic substances should be avoided.

应避免使人体接触有潜在毒性的物质。

(动词时为"expose A to B"，"使 A 接触到 B"。)

These special problems arise from the **use** of atomic energy **as** a source of power.

这些特殊问题是由把原子能用作能源而产生的。

(动词时为"use A as B"，"把 A 用作为 B"。)

The **definition** of an electric current **as** the flow of electric charge is familiar to all of us.

把电流定义为电荷的流动对我们大家都是熟悉的。

(动词时为"define A as B"，"把 A 定义为 B"。)

The energy radiated by the sun is due to the continuous **transformations** of hydrogen **into** helium in its interior.

太阳辐射的能量是由于在其内部连续地把氢转变成氦所引起的。

(动词时为"transform A into B"，"把 A 转变成 B"。)

The **separation** of aluminium **from** its ore was very difficult at that time.

把铝从矿石中提炼出来在当时是非常困难的。

（动词时为"separate A from B"，"把 A 从 B 中分离开来"。）

如果在上述情况下要表示该名词的逻辑主语的话，就要使用 by-短语，其模式如下：

名词（来自及物动词）＋of＋A［逻宾］＋（介词　B［逻状］）＋by＋C［逻主］

一般情况下"介词　B"这一部分是不出现的，若有的话也可以放在"by C"之后。例如：

What is the date of the **discovery** of America **by Columbus**?

哥伦布发现美洲是在哪一天？

The **formation** of the theory of relativity **by Einstein** is one of the most significant events of the 20<sup>th</sup> century.

爱因斯坦确立了相对论是二十世纪最重要的事件之一。

Niels Bohr，in 1913，first applied these ideas to the **emission** of light **by atoms**.

在 1913 年，尼尔斯·布尔首先把这些概念应用于原子发射光的方面。

The careful astronomical observations of Tycho Brahe and the brilliant **analysis** of his idea **by Johannes Kepler** illustrate very dramatically the contribution of accurate measurement to scientific progress.

泰乔·布雷厄仔细的天文观察以及由约汉尼斯·凯普勒对他的想法所作的精辟分析，非常明显地说明了精确的测量对科学进步的贡献。

The **discovery** of the periodic table **by Mendeleev** in 1869 suggested another way of estimating atomic masses.

门捷列夫于 1869 年发现了周期表使人们有了计算原子质量的另一种方法。

（"by Mendeleev"是"discovery"的逻辑主语，而"in 1869"为它的逻辑状语。）

## Ⅲ、几个小问题

（1）在"**Table X**"后不加任何标点符号，然后空一、二格写标题（可以用普通体，也可以用黑体），可以用正体，也可以用斜体，最后不加任何标点符号。在表格中每项标题的第一个字母要大写，其他实意单词前第一个字母可以大写，也可以小写（也有人全部字母用大写的）。例如：

**Table 9 – 2** Typical Specifications of Various Types of Capacitors

| Type | Usual capacitance range | Insulation resistance |
|---|---|---|

*Table 1.3 Average probability of error for repetition code*

| *Code Rate* | *Average Probability of Error* |
|---|---|

**TABLE 7.7** MAGNETIC DRUM CHARACTERISTICS

| CHARACTERISTICS | UNIVAC 1108 | 1964 ICL 1900 SERIES |
|---|---|---|

　（2）图示下方的说明可以用正体，也可用斜体。在"**Fig. X**""**Figure X**"或"**FIG X**"后面不加标点符号而是空一、二格写说明（一般用普通体，也有用黑体的），说明不论是句子还是短语其第一个字母要大写，其中的冠词可有可无，在结尾一定要加句号（即黑点）。例如：

**Fig. 7 – 39**　Balanced modulator used for demodulation of SSB.

**Fig. 24 – 1**　The leaf electroscope.

**Figure 3 – 3**　**Fermi-level conditions in a junction at equilibrium.**

***Figure 8.32***　*Block diagram of adaptive antenna array.*

**FIG 9 – 6**　An equivalent circuit of a capacitor.

***Fig. 24 – 5***　*F is the force on Q due to the upper charge q.*

　（3）在坐标图的情况下，轴上的说明中第一个单词的第一个字母要大写，其他均用小写；里边曲线部分的说明中也是第一个单词的第一个字母要大写，其他的字母用小写。

　（4）在方框图中，每个方框中名称的第一个单词的第一个字母要大写，其他的均用小写（也有人全用大写字母的）。

　（5）在"i.e."（＝that is"也就是"）后要加逗号；在"e.g."（＝for example"例如"）后要加逗号；在"viz."或"viz"（通常读成 namely"即"）后要加逗号；在"etc."（＝et cetera＝and so on"等等"）后要加逗号，但当处于句尾时不能再加一个句号（小黑点）了，其本身的小黑点就兼做句号了。

　（6）"**Fig. 1 and Fig. 2**"一般可以合在一起写成"**Figs. 1 and 2**"，其中的"s"

不能丢，这属于复数形式。（注意，如果是一个图中的几个小图，则"**Fig.**"往往不加"s"，看成单数，如：Fig. 5.2（a）and（b）**shows** this.）同样我们可以有"Eqs. 1 and 2"（式 1 和式 2）。

　　（7）如果两个名词不是合在一起的意思而仅仅是共用一下，则一般不用复数形式。如：This can be seen by comparing **Eq.（10.2）and（10.4）**.（这一点可以通过比较式（10.2）和式（10.4）看出来。）

# 参 考 文 献

[1]  秦荻辉. 学术英语写作典型错句详析 500 例. 西安：陕西科学技术出版社，2007.

[2]  秦荻辉，周正履，李滔等. 英语学术论文错误解析. 北京：石油工业出版社，2013.